797,885 Books
are available to read at

www.ForgottenBooks.com

Forgotten Books' App
Available for mobile, tablet & eReader

ISBN 978-1-330-28201-4
PIBN 10012289

This book is a reproduction of an important historical work. Forgotten Books uses state-of-the-art technology to digitally reconstruct the work, preserving the original format whilst repairing imperfections present in the aged copy. In rare cases, an imperfection in the original, such as a blemish or missing page, may be replicated in our edition. We do, however, repair the vast majority of imperfections successfully; any imperfections that remain are intentionally left to preserve the state of such historical works.

Forgotten Books is a registered trademark of FB &c Ltd.
Copyright © 2015 FB &c Ltd.
FB &c Ltd, Dalton House, 60 Windsor Avenue, London, SW19 2RR.
Company number 08720141. Registered in England and Wales.

For support please visit www.forgottenbooks.com

1 MONTH OF FREE READING

at

www.ForgottenBooks.com

By purchasing this book you are eligible for one month membership to ForgottenBooks.com, giving you unlimited access to our entire collection of over 700,000 titles via our web site and mobile apps.

To claim your free month visit:
www.forgottenbooks.com/free12289

* Offer is valid for 45 days from date of purchase. Terms and conditions apply.

English
Français
Deutsche
Italiano
Español
Português

www.forgottenbooks.com

Mythology Photography **Fiction** Fishing Christianity **Art** Cooking Essays Buddhism Freemasonry Medicine **Biology** Music **Ancient Egypt** Evolution Carpentry Physics Dance Geology **Mathematics** Fitness Shakespeare **Folklore** Yoga Marketing **Confidence** Immortality Biographies Poetry **Psychology** Witchcraft Electronics Chemistry History **Law** Accounting **Philosophy** Anthropology Alchemy Drama Quantum Mechanics Atheism Sexual Health **Ancient History Entrepreneurship** Languages Sport Paleontology Needlework Islam **Metaphysics** Investment Archaeology Parenting Statistics Criminology **Motivational**

THE ART AND CRAFT

OF

CABINET-MAKING.

WHITTAKER'S LIBRARY

OF

Arts, Sciences, Manufactures, and Industries.

Small Crown 8vo. cloth.   With many Illustrations.

'Messrs. Whittaker's valuable Series of Practical Manuals.'
Electrical Review.

THE ART AND CRAFT OF CABINET-MAKING.   By D. DENNING. With 219 Illustrations.

THE OPTICS OF PHOTOGRAPHY.   By J. TRAILL TAYLOR.   3s. 6d.

THE PRACTICAL TELEPHONE HANDBOOK OR GUIDE TO TELEPHONIC EXCHANGE.   By JOSEPH POOLE, Whitworth Scholar (1875), late Chief Electrician to the Lancashire and Cheshire Telephone Exchange Company.   300 pp.   With 227 Illustrations.   3s. 6d.

FIRST BOOK OF ELECTRICITY AND MAGNETISM.   By W. PERREN MAYCOCK, M. Inst. E.E.   With 85 Illustrations.   2s. 6d.

ELECTRIC LIGHT INSTALLATIONS AND THE MANAGEMENT OF ACCUMULATORS.   A Practical Handbook.   By Sir D. SALOMONS, Bart., M.A., Vice-President of the Institution of Electrical Engineers.   Sixth Revised and Enlarged Edition, with Illustrations.   6s.

ELECTRIC INSTRUMENT-MAKING FOR AMATEURS.   A Practical Handbook.   By S. R. BOTTONE.   Fourth Enlarged Edition.   With 60 Illustrations.   3s.

ELECTRIC BELLS, AND ALL ABOUT THEM.   A Practical Book for Practical Men.   Third Edition.   With 100 Illustrations   By S. R. BOTTONE.

ELECTRO-MOTORS: How Made and How Used.   By S. R. BOTTONE.   Second Edition, Revised and Enlarged.   3s.

ELECTRICITY IN OUR HOMES AND WORKSHOPS.   By SYDNEY F. WALKER, M.I.M.E., A.M.I.C E., M A.I.C.E.   Second Edition.   5s.

ELECTRICAL INFLUENCE MACHINES.   Containing a Full Account of their Historical Development, their Modern Forms, and their Practical Construction.   By J. GRAY, B.Sc.   4s. 6d.

PRACTICAL IRON FOUNDING.   By the Author of 'Pattern Making,' &c.   Illustrated with over 100 Engravings.   Cloth.   4s.

METAL TURNING.   By a Foreman Pattern Maker.   With 81 Illustrations.   4s.

THE ELECTRO-PLATERS HANDBOOK.   A Practical Manual for Amateurs and Young Students in Electro-Metallurgy.   By G. E. BONNEY.   With 61 Illustrations.   3s.

Complete Lists of Messrs. Whittaker's Practical Manuals sent Post Free on application.

London: Whittaker & Co., 2 White Hart Street, Paternoster Sq.

THE ART AND CRAFT

OF

CABINET-MAKING

A Practical Handbook

TO

THE CONSTRUCTION OF CABINET FURNITURE

THE USE OF TOOLS, FORMATION OF JOINTS,
HINTS ON DESIGNING AND SETTING OUT WORK,
VENEERING, ETC.

TOGETHER WITH

A REVIEW OF THE DEVELOPMENT OF FURNITURE

BY

DAVID DENNING

WITH TWO HUNDRED AND NINETEEN ILLUSTRATIONS

LONDON
WHITTAKER & CO., 2 WHITE HART STREET
PATERNOSTER SQUARE, AND
GEORGE BELL & SONS, YORK STREET, COVENT GARDEN
1891

LONDON
PRINTED BY STRANGEWAYS & SONS
Tower Street, Cambridge Circus, W.C.

PREFACE.

By way of preface it seems unnecessary to say much beyond stating that the intention is to supply amateurs and young professional cabinet-makers with a reliable guide to the construction of cabinet furniture. No attempt has been made to teach the thoroughly experienced artisan, and no new fads are advocated either in style or processes. The ordinary reliable methods of the workshop and nothing more are explained, and on this account the book will, no doubt, be of greater use to those for whom it is intended than if new theories, of 'construction as it ought to be,' according to many of those who presume to teach the skilled mechanic, had been advocated. It will, no doubt, have been observed by those who are interested in the subject that cabinet-making as distinguished from joinery has received scant attention, as with scarcely an exception the books professedly treating of the former only, have included much that pertains to the latter. Those who are practically acquainted with the manufacture of furniture will understand the reasons, which, however, it is unnecessary to explain here.

Those who may wish to have a further acquaintance with the construction of furniture, other than cabinet work, may be interested to know that it is in contemplation to treat of upholstery, fret-sawing, marquetry cutting, French and other polishing, &c., in other volumes.

It merely remains to explain that the chapter on brass-work has been, if I may so call it, edited by my friend, Mr. W. H. Bridge, who is well known in Birmingham and in furniture trade circles as an authority on the subject, and to thank him, as well as Mr. J. Thompson, for the kindly interest they, among others, have taken in this book, which has had the advantage of their criticism while in progress.

Thanks are also due to Messrs. Wm. Marples & Sons, Sheffield, for having kindly furnished illustrations of tools, which it will be noticed show their familiar trade mark, the shamrock. The Hibernia brand is sufficient guarantee as to the quality of any tools bearing it.

D. DENNING.

CONTENTS.

CHAPTER I.

INTRODUCTORY 1

Ordinary Joinery not Cabinet-making—Joiners' Furniture—Example of different forms of Tools—Subdivision of Labour in Furniture-making—Special Work of the Cabinet-maker—Subdivision of Cabinet-making into Branches—Causes of bad Furniture being made—Advantages of understanding general Cabinet Work—Amateurs' Work—Skill only to be gained by Practice—Necessity for Observing—Cabinet-making not frivolous Work—Thought required as much as Strength.

CHAPTER II.

REVIEW OF DEVELOPMENT OF FURNITURE . . 15

Sham Antique Furniture—Mistaken Ideas about Old Furniture—Furniture in Tudor Times—Development of Furniture—Carving on Old Furniture, and Restorers' Practices—Furniture of the Georgian Period—Introduction of Mahogany—Chippendale and Chippendale Furniture—Manwaring—Heppelwhite—Sheraton—Architects—Furniture of the first half of present Century—Modern 'Art' Furniture—Furniture Designers—Influence of Sir Chas. Eastlake—'Early English'—Recent Changes—Cause of Changes—Old and Modern Furniture contrasted—Superiority of good Modern Work over Old Furniture.

CHAPTER III.

FURNITURE WOODS 40

Mahogany — Cedar — Pencil Cedar — Oak — Walnut — Ash — Hungarian Ash — Rosewood — Birch — Beech — Satinwood — Pine — Pitch Pine — American Whitewood — Sequoia — Timbers occasionally used — Logs — Buying Timber — Measurements — Seasoning and Drying — Levelling Boards — Waste.

CHAPTER IV.

GLUE AND ITS PREPARATION 59

Frequent Use—Selecting Glue—Preparation—Employment—Preservation—Liquid Glues—Brush.

CONTENTS.

CHAPTER V.

NAILS 64

Screws—Sizes—Brass Screws—Brads—Wire or French Nails—Needle Points—Dowels—Dowel Plates—Glass Paper—Stopping.

CHAPTER VI.

TOOLS 70

Selection and Care—List of Tools—Saw Teeth—Panel Saw—Sharpening and Setting—Tenon Saw—Dovetail Saw—Bow Saw and Frame—Planes—Iron and Wooden Planes—Plane Irons—Jack Plane—Trying Plane—Smoothing Plane—Rabbet Plane—Plough—Old Woman's Tooth—Hollows and Rounds—Chisels—Gouges—Spokeshave—Gimlets—Bradawls—Brace and Bits—Gauge for Bits—Screwdrivers—Marking, Cutting, and Mortise Gauges—Pincers and Pliers—Square—Bevel—Hammer—Mallet—Punch—Compasses—Rule—Scraper—Scraper Sharpener—Marking Awl—Grindstone—Cork—Handscrews—Holdfast—Cramps—Files—Dowel Plates.

CHAPTER VII.

WOODEN APPLIANCES MADE BY THE USER . 106

Cramp—Extemporised Cramps—Shooting Boards—Mitre Shooting Boards—Mitre Block—Mitre Box—Square—Straight Edge—Winding Strips—Scratch or Router—Benches—Tool-Chest.

CHAPTER VIII.

GRINDING AND SHARPENING TOOLS 127

Angles of Cutting Edges—Workshop Practice—Grinding and Sharpening Edge Tools—How to Hold Them—Sharpening Scrapers.

CHAPTER IX.

GENERAL DIRECTIONS ON THE USE OF TOOLS 131

Sawing—Planing—Scraping—Boring with Brace and Bits—Use of Winding Sticks—Circular Saw—Lathes—Fret Machine.

CHAPTER X.

JOINTS 146

Squaring up — Edge Joints — Plain Gluing — Dowelling — Tonguing — Plain Dovetailing — Lap Dovetailing — Mitred Dovetailing — Bearers — Keyed Corners — Mortises and Tenons — Dowelled Frames — Halving.

CHAPTER XI.

ECORATIVE AND MINOR STRUCTURAL DETAILS 165

Lining-up — Rabbeting — Bevelled-edge Panels — Cross Grooving — Stop Chamfering — V Grooves — Beaded Edges — Stopped Beads — Flutes — Inlaid Stringing — Mouldings — Panels — Facing.

CHAPTER XII.

CONSTRUCTION OF PARTS 183

Drawers — Doors — Cornices — Plinths.

CHAPTER XIII.

GLASS IN FURNITURE 196

Sheet Glass — Plate Glass — Purchasing — Flaws — Bevelling — Silvering — Measuring — Fixing.

CHAPTER XIV.

DRAWING AND DESIGNING 209

Considerations for Guidance in setting out Work — Miniature Designs — Working Drawings.

CHAPTER XV.

VENEERING 221

Objections considered — Burr Veneers — Saw-cut and Knife-cut Veneers — Laying with Caul — Wooden Cauls — Metal Cauls — Care and Preparation of Veneers — Preparation of Wood for Veneering on — Light-coloured Veneers — Cleaning up Veneered Work — Laying with Hammer — Veneering Hammer — Blisters — Veneering on End Grain — Inlaid Veneers — Veneering Curved Surfaces.

CHAPTER XVI.

CABINET BRASS-WORK 239

Till Locks—Cut Cupboard Locks—Box Locks—Desk Locks Straight Cupboard Locks—Wardrobe Locks—Nettlefold's Piano Lock—Spring Catches—Flush Bolts—Socket Castors —Screw Castors—Pin Castors—Castor Rims—Dining-table and Pivot Castors—Iron-plate Castors—Ball Castors—Wright's Ball Castor—Paw Castors—Butt Hinges—Back-flap Hinges Card-table Hinges—Desk Hinges—Screen Hinges—Centre Hinges—Piano Hinges—Hinge Plates—Escutcheon Plates and Thread Escutcheons—Brass Handles.

CHAPTER XVII.

CONSTRUCTION—TABLES 252

General Advice—Simple Fancy Table—Small Table with Shelf Below—Octagon Table with Spindled Rails—Square Tapered Legs—Small Round Table—Common Kitchen Table—Leg Writing Table—Table with Flaps—Brackets for supporting Flaps—Sutherland Table—Double Sutherland Table—Rule Joint—Card Tables—Dining Tables.

CHAPTER XVIII.

BEDROOM FURNITURE 279

Plain Hanging Wardrobe—Small Wardrobe with Drawer—Wardrobe with Straight Ends—Beaconsfield Wardrobe—Six-ft. Wardrobe with Long Trays and Drawers—Short Trays and Drawers—Fittings—Toilet Table Glass—Dressing Chests—Washstands—Pedestal Cupboard—Pedestal Toilets.

CHAPTER XIX.

LIBRARY AND OFFICE FURNITURE 301

Pedestal Writing-table—Double ditto—Desk Slopes—Register Writing-tables—Cylinder Fall-tables—Old Bureau—Dwarf Bookcases—Secretary Bookcase—Nests of Pigeon-holes.

CHAPTER XX.

SIDEBOARDS AND CABINETS 313

Ordinary Arrangement of Sideboards—Fixing of Back—Cabinets —Music Cabinets.

LIST OF ILLUSTRATIONS.

FIG.
1. Handsaw.
2. Size and Shape of Saw Teeth.
3. Rip-saw Teeth.
4. Tenon Saw.
5. Bow Saw.
6. Jack Plane.
7. Angle for Plane Iron Edge.
8. Unsuitable Edge.
9. Plane Iron.
10. Edge requiring Regrinding.
11. Curve of Jack Plane Iron.
12. Trying Plane.
13. Smoothing Plane.
14. Rabbet Plane.
15. Plough.
16. Old Woman's Tooth.
17. Paring Chisel.
18. Firmer ,,
19. Mortise ,,
20. Gouge.
21. Spokeshave.
22. Shell Gimlet.
23. Twist ,,
24. Bradawl.
25. Brace.
26. Centre Bit.
27. Twist Bit and Gauge.
28. Ordinary Screwdriver.
29. Marking Gauge.
30. Mortise ,,
31. Square.
32. Try and Mitre Square.
33. Sliding Bevel.
34. Hammer.
35. Wing Compasses.
36. Marking Awl.
37. Marking Chisel.
38. Oilstone in Case.

FIG.
39. Hand-screws.
40. Bench Holdfast.
41. Iron Cramp.
42. Box and Tap for Wood-screws.
43. Wooden Cramp.
44. Bar for Screw-block.
45. Screw-block to fit Bar.
46. Bar showing Groove.
47. Sliding-block showing Tongue.
48. Iron Link.
49. Improvised Cramp.
50. ,, ,,
51. Shooting Board.
52. ,, ,, or Bench Hook.
53. Mitre Shoot.
54. Simple Mitre Shoot.
55. Mitre Block.
56. Mitre Box.
57. Testing Square.
58. Scratch or Router.
59. Cutter for Bead.
60. Bead.
61. Rows of Beads.
62. Beads with Hollow between.
63, 64. Cutters for Hollows.
65, 66. Scratch for Chamfering.
67. Cheap Wooden Bench.
68. Iron Standard Bench.
69. German Bench.
70. Badly Ground Edge.
71. Properly ,, ,,
72. Scraper Edge, sharp.
73. ,, ,, round.
74. Winding Sticks on Board.
75. ,, ,, showing Board in Winding.
76. Circular Saw with Dovetailing appliance.

LIST OF ILLUSTRATIONS.

FIG.
77. Lathe.
78, 79. Wood glued on to strengthen joint.
80. Tongued joint.
81. ,,   ,,
82. Plain Dovetails—separate parts.
83.   ,,   ,,   together.
84. Ditto, with Badly-formed Pins.
85. Lap Dovetail.
86. Formation of Mitre-lap Dovetails.
87. Dovetailed Top-bearers.
88. Dovetail on End.
89.   ,,   ,,
90.   ,,   Stopped Back.
91. Mitred and Keyed Corner.
92. Mortise and Tenon.
93.   ,,   ,,
94. Double Tenon.
95. Tenon with Haunch.
96. Tenons on End of Shelf, &c.
97. Foxed Tenon.
98. Dowel-jointed Frame.
99. Halved Corner Joint.
100.   ,,   Joint.
101.   ,,   Mitred Joint.
102-5. Sections of Lined-up Tops.
106. Top with Lining.
107. Rabbet.
108. Stopped Rabbet.
109. Panel with Sunk Bevelled Edges.
110. Section of Board notched for grooving with Saw.
111. Stop Chamfered Edges.
112. Guide for cutting Chamfer Stops.
113. Stop Chamfering Tool.
114. Door with Chamfered Edges, Frame, and V-grooved panel.
115. Section of V Groove.
116-119. Moulded Edges.
120-122. Stopped Beads.
123. Flutes.
124. Inlaid Bandings.
125. Ovolo Moulding.
126. Thumb   ,,

FIG.
127. Ogee Moulding.
128. Formation of Ovolo.
129. Moulding.
130.   ,,   from Solid.
131.   ,,   ,,   Board.
132. Alternative Form.
133. Flush Panel Rabbeted.
134. Drawer Side.
135. Fitting of Drawer Bottom.
136. Drawer Side with Bottom, &c.
137. Alternative Fitting for Drawer Bottom.
138. Munting for Bottom.
139. Ditto, showing Back.
140. Fastening of Drawer Runner.
141, 142. Parts of Drawer Fittings.
143. Drawer Front with Moulding.
144. Ditto—Section.
145. Panel Rabbeted in Frame.
146. Rabbet formed with Moulding.
147. Mitred Mouldings worked on Frame.
148. Moulding sunk in Rabbet.
149. End of Stile showing Rabbet.
149a. Plinth or Cornice with fastenings.
150. Cornice Moulding.
151. Lining of Plinth.
152-154. Shapes of Bevelled Glass.
155. Glass Fitting close to Frame.
156. Glass not Fitting close to Frame.
157. Bevelled Glass Blocked in.
158. Transparent Bevelled Glass Fitted.
159. Overmantel.
160.   ,,   Front Elevation.
161.   ,,   End Elevation.
162.   ,,   Working Drawing and Scale.
163. Overmantel, Working Drawing of End.
164. Section of Wood showing Heart Side.
165. Veneering Hammer.
166. Diagram showing Course of Veneering Hammer.

LIST OF ILLUSTRATIONS.

FIG.
167. Small Table.
168. Screw through Rail.
169. Small Table with Tray.
170. Fitting of Legs.
171. Table with Spindled Rail.
172. Fitting of Spindles.
173. Square-legged Table.
174. Foot of Square Leg.
175. Small Fancy Table.
176. Round   ,,
177. Square-legged Common Table.
178. Flap Table.
179. Folding Bracket.
180. Knuckle Joint.
181.   ,,   ,, Setting Out.
182.   ,,   ,, Shaping of Parts.
183. Finger Joint.
184. Swing Rail for Supporting Flap.
185. Sutherland Table.
186. Double Sutherland Table.
187. Table with Swinging Legs.
188. Rule Joint.
189.   ,,   Setting out for.
190. Folding Card-table Top.
191. Extending Dining-table Frame.
192. Construction of Slides.
193. Dovetailed Slide.
194. Small Hanging Wardrobe.
195. Hanging Wardrobe with Drawer.

FIG.
196. Wardrobe with Straight Ends.
197. 'Beaconsfield' Wardrobe.
198. Six-ft. Wardrobe.
199. Interior of Three-door Wardrobe.
200. Toilet Table with Glass.
201.   ,,   ,,
202. Dressing-chest and Glass.
203. Washstand, with Single Row of Tiles in Back.
204. Washstand, with Pedestal Cupboard.
205. Pedestal Washstand.
206. Pedestal Toilet-table.
207. Pedestal Writing-table.
208. Desk Slope.
209. Pedestal Register Writing-desk.
210. Upper Portion of Half Register.
211. Top of Cylinder-fall Table.
212. Method of Hanging Fall.
213. Top of Writing-desk with Tambour Fall.
214. Bureau.
215. Dwarf Bookcase.
216. Rack for Movable Shelves.
217. Dwarf Bookcase with Break Front.
218. Secretary Bookcase.
219. Music Cabinet.

THE ART AND CRAFT OF CABINET-MAKING.

ANNOUNCEMENT.

HOWEVER lucid instructions may be it is impossible in such a complex subject as cabinet-making to foresee every difficulty which may occur to the novice or to give directions which shall cover and apply in every instance. Unless the design and specification are alike each piece of furniture differs in some detail of construction from others, and on each point of divergence the beginner may, even if he does not meet with a serious difficulty, be in some doubt as to the best or correct method. To meet such cases the publishers have pleasure in stating that they have arranged with the author of this book for those who wish to do so to avail themselves of his advice and experience.

This is no doubt an innovation, but there are few whose information on any craft is derived mainly or even partially from books who have not felt that it would be an advantage to avail themselves of skilled advice. The want is no doubt partially supplied by some technical periodicals, but from the interval which must necessarily elapse before the inquirer's answer can be published, much, and often valuable, time is necessarily wasted. By direct correspondence this can be avoided, and there are no doubt many who will gladly avail themselves of the facilities now afforded of quickly obtaining reliable advice and guidance. In addition to helping those in difficulty, it will be of interest to many to know that they can be supplied with either small or working drawings, and that in the event of these being made by themselves they can submit them for rectification if necessary. By this means loss caused by mistakes, which may not be discovered till the work is being proceeded with, can be avoided, and beginners will have the satisfaction of knowing that they are working on correct lines.

ANNOUNCEMENT.

It is only natural that the author cannot undertake this work or devote time and attention to it without remuneration, and it is quite impossible to give a scale of fees which will be applicable in every instance. The fees will be regulated by the information required; or, in the case of drawings, by the amount of work, and will be kept as low as they consistently can be in order that all may avail themselves of the privileges offered.

Those who wish to avail themselves of the Author's aid must proceed as follows :-

State as fully as possible the cause of difficulty and information wanted, or if drawings are to be revised, send them. As soon as possible Mr. Denning will state his fee, and on receipt of this will supply what is wanted.

All letters for him must be enclosed to the publishers with a stamped envelope directed merely Mr. Denning, leaving space for his address to be filled in by them. A stamped addressed envelope must be enclosed for Mr. Denning's reply, which will be sent direct to the inquirer.

Every care will be taken, but no responsibility can be undertaken either by the publishers or by the author in the event of loss or mistake.

Mr. Denning may be consulted in any part of the country on payment of adequate fees and travelling expenses, so that those who wish to fit up or arrange workshops or rooms may have his advice. He will also undertake to recommend unusual or new tools which may be specially applicable to any particular class of work. Indeed, to sum up it may be said that the author is willing to act as adviser on any matter connected with furniture, and in making this arrangement the publishers feel confident that they are studying the best interests of the readers who may desire to improve themselves.

It may also be said for the benefit of those who wish to take personal lessons in cabinet-making, that in most of the larger towns Mr. Denning can give the names of skilled artisans who will do what is necessary.

(Signed)   WHITTAKER & CO

THE ART AND CRAFT OF CABINET-MAKING.

CHAPTER I.

INTRODUCTORY.

Ordinary Joinery not Cabinet-making—Joiners' Furniture—Example of different forms of Tools—Subdivision of Labour in Furniture-making—Special Work of the Cabinet-maker—Subdivision of Cabinet-making into Branches—Causes of bad Furniture being made—Advantages of understanding general Cabinet Work—Amateurs' Work—Skill only to be gained by Practice—Necessity for Observing—Cabinet-making not frivolous Work—Thought required as much as Strength.

As there are apparently many popular misconceptions on the subject, it may be well at the beginning to make some attempt to explain the scope of the cabinet-maker's work, and show wherein it is distinct and separate from kindred or allied crafts. In ordinary conventional language, the cabinet-maker is one who either sells or makes furniture of all kinds, or perhaps does both. The present concern, however, is not with him as a dealer, but as an artisan, a craftsman who makes ordinary wooden domestic furniture. This clears the way somewhat, and it will be well to consider what, in the capacity of cabinet-maker, one is expected to do, though it must be admitted it is not altogether an easy matter to define the limits of the craft without occupying an undue amount of space. It is, however, necessary that the worker should have a tolerably clear conception of what is expected of him. Broadly speaking, the cabinet-maker—only as a craftsman, mind—has nothing to do with upholstery, nor is he the same individual as the polisher who finishes the work. The cabinet-maker confines his attention to the woodwork of furniture. He fashions and forms the wood, fastening it together, but has little or nothing to do with other materials, except so far as they are necessary for construction. He is, therefore, a joiner; but to confound joinery or carpentry with cabinet-making is altogether a mistake, though one commonly made. Because a man is a cabinet-maker, and consequently works with wood, he is supposed to be competent to do anything required in the joinery way. From his familiarity with the material chiefly used he may have more aptitude than one quite unaccustomed to wood or wood-working tools in carpentering generally, but neither a carpenter nor a cabinet-maker is so much at home in the work of the other as in his own special line. The joiner is principally concerned with large work and with comparatively soft woods, while the cabinet-maker is, from the nature of the articles he makes, occupied principally with small constructions in the choicer and harder woods. Without at all seeking to decry the skill of the joiner, it may be admitted that cabinet-making is finer work—fine joinery in fact—and nothing else. I am quite aware that many joiners can, and do, make furniture, and there are not wanting those who consider that articles of furniture are of better quality when made by the joiner or carpenter. Joiners' furniture is, however, lacking in the finish which is imparted to it by a good cabinet-maker, unless, indeed, the joiner has had some training in the special work of the latter, or, in other words, has learned cabinet-making. As for the quality, by which is meant the superiority in solidity and general construction, it amounts to little more than crudeness of work and methods. Those who think otherwise have generally little or no practical knowledge of the subject either from an artisan or a commercial point of view, but are led away by entirely false notions of so-called art. This, however, is a subject which I shall have occasion to refer to in a subsequent chapter, so that for the present it may be dismissed. I think there can be little doubt that at one time the cabinet-maker and the joiner were one and the same person, the two crafts having drifted apart owing to the special features of each having become developed, partly from ordinary conditions of business, and partly from force of individual circumstances or preference, for it is no uncommon thing to find that a joiner has become exclusively a cabinet-maker, and *vice versa*. This is not surprising, for the tools, with certain modifications in some cases, are identical, and there is no reason why one who can make, say, a door of a room, should not be able to make a similar part for a bookcase, sideboard, or anything else. If accustomed to working in pine he might, naturally would, feel somewhat awkward at first with a piece of fine, hard, figury 'Spanish,' but that would soon be overcome. Then there are the various little details in connexion with cabinet-making which are different from the methods practised in joinery. For instance, our friend the joiner, working, as has been said, principally in pine, would find that the fillister is not so suitable as the rabbet-plane for making a rabbet in hard-wood furniture, a fact which the trained cabinet-maker is aware of, and so very probably does not even possess a fillister, the use of which would be somewhat inconvenient to him on that account even in 'deal' work. If, therefore, the novice does not find mentioned in the following pages some tool which he is aware of, and perhaps has in

INTRODUCTORY.

a more or less hazy way regarded as indispensable, he may understand that though suitable for joinery, it is not generally used by the cabinet-maker, or, if the expression be preferred, by the joiner who exclusively devotes his attention to the construction of furniture. The same may be said about joints, for though there are many of these which are of general adoption, some, such as the scarf, are needless in cabinet-making, while others little used in joinery or building construction are of frequent application in it. On such minor details the distinction between cabinet-making and joinery chiefly consists, so that a good joiner has little difficulty in becoming equally facile as a cabinet-maker, while the latter finds the transition equally easy.

So far the cabinet-making craft has been regarded in its widest application, for in practice it is considerably more restricted than anything which has been said would imply. Upholstery and polishing, it has been already stated, are distinct from cabinet-making, but the fact that this is again subdivided must not be lost sight of. The cabinet-maker of modern times may be a very competent individual — or, according to some, an incompetent one — but he does not profess to be either a turner, an inlayer or marquetry cutter, a fretsawyer, or a carver. What, then, does he do? some may be inclined to ask. Well, it may be answered for the enlightenment of such people, he makes up the things, and he has plenty to occupy him in so doing. Suppose we take, by way of illustration, a sideboard or a cabinet in which there are turned columns, carved parts, marquetry panels, and one or two bits of fretwork. In addition to the construction of the article, the cabinet-maker would get out the square pieces to be turned, the pieces to be carved, lay the marquetry veneers on the panels, and prepare the pieces for the frets. To a certain extent the turner, the carver, the fretsawyer, and the marquetry

INTRODUCTORY.

cutter are subsidiary to him: their work is decorative, his constructive. The art of the cabinet-maker does not consist in decorating his work, but in making it, putting the parts together properly, substantially, and with neatness. It is, therefore, quite possible for one to excel as a cabinet-maker without having any practical skill in the more purely decorative branches of woodwork. It is quite true, however, that a piece of furniture should be decorative and ornamental in itself, but it by no means follows that carving and other adornments of a like kind are essential to beauty. Suitability to its intended purpose and accurate workmanship are of far more importance, and—though perhaps rather prematurely—I cannot refrain from impressing on the novice that plainness and ugliness are not synonymous, for it will at any rate show him that to be a really efficient cabinet-maker he need not be an 'Admirable Crichton' in woodwork. Here and there one may meet with a man who can do all that is required in making any piece of furniture, but he is the exception proving the rule. This is, that each devotes himself to a special branch, with the result that he becomes expert in it, instead of being merely fairly good all round.

There is, of course, a good deal to be said both for and against this subdivision of labour, which certainly does not meet with the approval of those who, ignoring the conditions of modern trade, would have us revert to the custom at one time prevalent, or supposed to have been so, of the same workman making and finishing a thing outright. It is not, however, unreasonable to assume that even in the days of auld lang syne the joiner or cabinet-maker would not disdain the aid of his fellow-craftsman who had made himself more than ordinarily expert in carving or other decorative work, to adorn his own crude and plain construction. There is little or no evidence to show that old-time workers did everything themselves instead of getting those specially skilled in particular branches to help them, when they could. We must, of course, remember that cabinet-making, as understood nowadays, is a comparatively modern development, and that our present furniture is very different from that of a couple of hundred years ago. 'Yes,' says Cynic, 'it undoubtedly is, for then it was sound, substantial, and artistic, while now it is——' I will finish the sentence by adding, 'just what cynic and other art-cranks like to pay for.'

That much, very much, bad furniture is made, of the jerriest construction and of the poorest materials, cannot be denied, but such defective work is to attributed only in a very small degree to the subdivision of labour, if, indeed, this is the cause of any of it. I would instead be inclined to say that the demand for cheap, or, rather, low-priced furniture, has originated much of the excessive subdivision which exists, and so great is the competition that the tendency is increasing in this direction. Many of the men called by courtesy, or popularly considered, cabinet-makers, are not so; they are specialists who make one article or class of article only, and are entirely ignorant both of the construction of other pieces of furniture and of the general principles of cabinet-making. This is particularly the case in London, whence a very large proportion of the furniture used throughout Great Britain emanates. Made under trade conditions there, much of it is utterly bad; so bad, that were the material any better than it is, one would be inclined to look on it as a waste of wood. Nevertheless, there is considerable sale for such rubbish, the manufacture of which would otherwise soon cease, and we find that the perverted ingenuity of many of the so-called cabinet-makers has enabled them to put together with the smallest amount of labour, and in utter defiance of all constructive considerations, beyond that

INTRODUCTORY.

of low price, things which to those who have no more than a superficial or ordinary acquaintance with furniture look and seem all right, while new. As soon as they get used, their quality is clear enough even to the most unsophisticated. If, then, furniture of the class alluded to is to cease from occupying the prominent place it does, the purchaser must be willing to pay a reasonable price for good work. There is no difficulty in obtaining it. Some of that done by special makers even is good, especially considering the conditions under which it is made. In case it may be thought that I am unduly severe on London cabinet-makers, it may be said that no one is more willing to admit the skill of many of them, perhaps even of the great majority, although there are far too many who might well be spared in the interests of good furniture production. Even among the small classes into which cabinet-makers are divided, there are some who are good all-round men, but the inevitable tendency among those who make only one kind of thing is towards inefficiency in making anything else. They get out of the run, as it were, of being general cabinet-makers, and instead become limited in their sphere. We thus find men who are makers of sideboards, wardrobes, toilet-tables, dining-tables, writing-tables, chests of drawers, bookcases, as well as the miscellaneous odds and ends known as fancy cabinet articles. Of course, in their own special lines these men are generally expert, and can work much more speedily, and therefore more cheaply, than those cabinet-makers who are able to undertake to make any article of furniture. Such specialism is not, however, of advantage to the worker, and cannot be considered conducive to the development of skill. A specialist rarely gets far from his own groove, whereas the man who has a good general knowledge of construction can have no great difficulty in turning his attention to any class of work which may demand it. From a trade or commercial point of view, subdivision of the general cabinet-making industry into a number of smaller ones may be a necessity, but I strongly urge the beginner to endeavour, as far as possible, to qualify himself for general work. If a 'carcase' worker, he will, at any rate, be none the worse off for being able to make a dining-table or anything else. Although London has been mentioned as the headquarters of specialism, the tendency towards it is more or less evident in large towns, though not to the same extent. From the nature of things it cannot be, so that on the whole the provincial cabinet-maker may almost be regarded as a more competent all-round worker than his metropolitan *confrère*. I am afraid this view of the matter may not be agreeable to all, but it is the result of many years' observation, and I may say I have no desire to exalt the country worker at the expense of the other. At the same time, I certainly cannot agree with so many Londoners who assume that the headquarters of cabinet-making skill are in and about the Curtain Road. For cleverness in making up cheap work its neighbourhood has the pre-eminence, for no cabinet-maker elsewhere could pretend to turn out such furniture. It will be understood that only a few of the subdivisions have been named, for there is hardly an article of furniture which has not its special makers. They make more or less in quantities, or, perhaps it will be better to say, as many of the masters are only in a small way of business, never make anything out of their ordinary run, and it is no uncommon thing to find methods adopted which are not generally practised, and cannot be spoken of with praise. It is, in fact, an approach to machine work; good, sometimes, in itself when not abused and within limits, but not suitable for the amateur, nor beneficial to the individual worker, nor conducive to general excellence.

The amateur is not bound by the same restrictions as the professional worker, and there is no reason why he should not make himself a good general cabinet-maker. He certainly labours under the disadvantage, and it must be confessed it is no small one, of not having sufficient opportunities of practice, but on the other hand he is able to take his own time in making anything. It is entirely his own fault if he passes a piece of work just because it will do, instead of making it as nearly perfect as possible. He can stand over a job as long as he likes without feeling that he is wasting his employer's time, or his own if he is doing piece-work. In fact, he is his own master, and should rather value excellence of work than consider how soon he can get a thing finished. Of course, in theory, the professional should do likewise, but in practice he seldom can, and sometimes does not want to. Still, I must say that the majority of those cabinet-makers whom I know do take a positive pride in their work, however much some people who know no better may be inclined to sneer at the British workman. Were it not so this book would probably never have been contemplated or written as a humble contribution to the literature of cabinet-making, a subject on which as distinct from joinery and carpentering work remarkably little has been written from a practical point of view.

It is, of course, impossible within a moderate compass to tell everything that might be told concerning cabinet-making, for among experts there are many ways of doing the same thing, and it is not reasonable to suppose that any one can possibly be acquainted with them all. Even if everything connected with cabinet-making could be told in detail the result would probably tend rather to perplex than help the novice, for whom this book is chiefly intended, whether he be amateur or professional.

The amateur, it is hoped, will find all necessary matters so fully explained that he will, if he follows the directions, be able to make any ordinary piece of furniture in a manner satisfactory not only to himself but to those who are competent to form an opinion of the quality of the work. To the professional cabinet-maker, whose experience may be limited, such a book as the present will doubtless be useful as showing him on what lines to proceed in making furniture other than he has been accustomed to. This last sentence will serve to explain that the construction and modes of procedure advocated are not of an amateurish character, prepared in a manner which purports to be simple for the benefit of the amateur worker. They are, on the contrary, thoroughly practical, and such as would not be objected to in any cabinet-maker's workshop in this country. Perhaps the amateur might have preferred to be told some easy way of making everything—to be shown, as it were, a royal road to cabinet-making in its entirety. I may as well say openly that I am unable to do this. The thing is impossible, for there is no way other than by downright hard work and perseverance of acquiring manipulative skill. The novice who thinks that he can right off make some piece of furniture in the most beautiful manner, nay, even that he can saw straight, or plane up a board true and smooth, will find himself grievously disappointed. He will find himself helped in the right course, or shall I not rather say started on it, by the hints in the following pages; but his progress must depend entirely on himself and his own aptitude for mechanical work. The same remark may be taken to himself by the apprentice or young journeyman who, if he follows his craft with enthusiasm, will find that he is never done learning. At present, that is while he is a young man, he may think he knows all about everything connected with the practical part of cabinet-making. He is, perhaps, on a level with his shopmates, and has learned all they can tell him, or fancies he has, which so far as his self-satisfaction is concerned is much the same thing. One day a new hand comes along and uses a tool, very likely some contrivance of his own, or does something in a different way from the usual one practised in that particular shop. Ah, young man, there is something new already for you to learn, and if you are wise you will at the same time have learned an even more useful lesson, viz., that there are still some things connected with your calling which you are unacquainted with. When you begin to find that out, you have made a distinct advance on your way to become a really efficient practical worker. I have known many youngsters who for a time thought there was very little more to be learned, but they found out their mistake sooner or later. I know I did, and I do not think I have ever known a middle-aged man who would pretend to know all about cabinet-making. There is always something fresh to be learned, for every man has some peculiarities in his methods, and the novice certainly cannot afford not to notice these and profit by them when he can. There are also new tools constantly coming forward, and sometimes one that is better than those which have hitherto been used is introduced, but as a rule one likes to use those he has become familiar with. Then there are modifications in the style of furniture, for this, like dress, has its fashions; and all must be noted by him who would excel.

The foregoing will to some extent serve as an answer to the question, How long does it take to learn cabinet-making? but in addition it must be stated that a novice of ordinary intelligence and dexterity will soon become fairly efficient as a mechanic, and be able to make plain and simple constructions. When he can do this he will, with care, gradually make progress till ere long he finds that he can make anything. He will, however, never be able to dispense with the need of care and accuracy in working, that is to say, the work can never become play. To the amateur it may become a pastime, but it can never be a frivolous one. With the artisan who pursues the work as a means of livelihood the case is somewhat different, 'no faithful workman finds his task a pastime.'

Perhaps something may be said about the advantages of cabinet-making as a hobby, though really very few remarks can be necessary. It may be taken for granted that the days have gone by when the amateur workman was a rarity, for their number now is legion, and more than one periodical caters specially for them. There can be no doubt that an intelligent pursuit of any mechanical work is of benefit to the worker, not only as a change from perhaps more sedentary occupation, but as possessing intrinsic interest. The following quotation from one of our most sensible technical authors, Chas. G. Leland, may possibly put the matter in a new light to some. In his *Manual of Wood-Carving* he says, 'even a very little frequent employment of the mind, inventing and planning, no matter at what, stimulates *all* the mental faculties.' We are rather too accustomed to regard handwork as almost independent of the head, and considerably lower in the scale. When the former degenerates into purely mechanical operations there may be little ennobling in it, but in any work involving more than labour or physical strength this cannot be the case. In cabinet-making the brain and hand must work together. There is constant opportunity for thought as well as for mere manipulative skill, and therefore it is worthy of earnest study both by the amateur and the professional craftsman. The words of Carlyle's creation Teufelsdröckh are as applicable to cabinet-making as to the subject in connexion with which they were originally uttered, for 'neither' in it 'does man proceed by mere accident, but the hand is ever guided on by mysterious operations of the mind.' The mind alone, though, will not give skill to the artificer; he must learn to use his hands, and, as has been said, practice is necessary. Theoretical knowledge can be acquired from books, practical skill can only be got by work. Even the former is better than none at all, for it will to some extent enable a man to discriminate between good and bad workmanship. Herein lies my answer to those who object to the general public being informed as to methods of work in any craft. The honest worker has nothing to fear from knowledge being widely diffused, for those who are able to judge of the quality of his work will be best able to appreciate its value. As for the other kind of worker, I have no sympathy with him unless he has been driven to make rubbish by sheer necessity, in which case he is to be pitied, and the best help that can be given is by showing as plainly as possible how well-constructed furniture ought to be made.

Perhaps, before closing this chapter, it may be desirable to state that chair-work is generally a distinct branch of trade, I mean even so far as the woodwork is concerned. The cabinet-maker may be able to make a chair frame, much in the same way that he could make anything required of wood, but, as a rule, he does not do so. The chair-maker and the cabinet-maker are distinct craftsmen, so that the construction of all kinds of frames for stuffing and upholstering occupies no place in the present volume, but will be treated of in another of the same series.

To those who have had any training in the workshop, many of the matters mentioned in the following pages may seem trivial and of too elementary a character. Should any readers think so, I must beg of them to remember that their advantages are not shared by all, and that a large number may have had no opportunity of acquiring practical knowledge. For their benefit, then, in order that they may profit and make use of the more advanced remarks, it has been deemed advisable to give prominence to details which are matters of common knowledge even to those who have worked only a few months under the competent guidance of practical cabinet-makers.

CHAPTER II.

REVIEW OF DEVELOPMENT OF FURNITURE.

Sham Antique Furniture—Mistaken Ideas about Old Furniture—Furniture in Tudor Times—Development of Furniture—Carving on Old Furniture, and Restorers' Practices—Furniture of the Georgian Period—Introduction of Mahogany—Chippendale and Chippendale Furniture—Manwaring—Heppelwhite—Sheraton—Architects—Furniture of the first half of present Century—Modern 'Art' Furniture—Furniture Designers Influence of Sir Chas. Eastlake—'Early English'—Recent Changes Cause of Changes—Old and Modern Furniture contrasted—Superiority of good Modern Work over Old Furniture.

THE important position occupied by furniture at the present day seems such a matter of course in the appointments of our homes that one seldom stops to consider that cabinet-making, as we understand it, is a craft of comparatively recent origin. We know, of course, that some things are old-fashioned, but beyond them all is chaos, so far as furniture is concerned, unless, indeed, we are aware of certain contrivances made of oak, usually more or less carved, and vaguely known as 'antique.' It is marvellous how antique some of these things are, in the opinion of their owners, founded, more likely than not, on the assertion of the broker or curiosity dealer who sold them the valuable articles. This gentleman, by the way, is often very accommodating, and will fix a date to suit his customer. We thus find 'grandfather' clocks of a date long anterior to that of the great discovery of Huyghens. Wonderful pieces of mechanism, those old clocks. 'Three or four hundred years old, my dear sir, and keeps better time than any other in the house.' The latter part of the assertion may be true, but the former cannot be accepted, if one pays any regard to the date of the application of the pendulum to time-keeping purposes. No, my friend, the age of your old clock is probably expressed by two figures, although it may run into three, the first of them being a 1. Not by any chance can it be much over 200 years old, and even then it must be a very early specimen, a great rarity. Clocks, or rather the clock cases, it must be remembered, were one time made by cabinet-makers, or they would not be mentioned here.

It is almost a pity to disturb the equanimity of the dear old gentleman who shows us with pride a sofa which King Henry VIII. is said to have used, and it is, moreover, a ticklish matter to sweep away fancy with facts. The general style of the thing shows it to have been made early in the present century, but as that is not convincing enough, it has to be pointed out that the wood is mahogany, which was unknown in this country till within the last decade of the sixteenth century, and was not used by either royalty or populace till more than a hundred years later. History tells us King Henry died considerably before the dates mentioned.

Then there is our friend the popular actor, who one time acquired a veritable curiosity in the antique furniture line, nothing less than a genuine (?) old sideboard, oak covered with carving, with plate-glass back, cellarette and all complete, a modern sideboard in antique garb. Goodness only knows what the wonderful history attached to it was, but it did credit to the imagination of some one, even if it did not proclaim his veracity. The chief indictment against that piece of furniture was that our Elizabethan ancestors did not have sideboards, except in the literal sense of side boards, boards at the side of the room, and they never, for the best of all reasons, had large pieces of silvered plated glass above them.

The word 'Chippendale' is responsible for much

disappointment to those who collect or set store by old furniture without knowing anything about it. There seems something in the name which renders it more adaptable to old pieces of furniture than any other. Witness the fortunate owner of some Chippendale chairs 'black with age, two or three hundred years old, and made of mahogany.' I never saw those chairs, but there is something in the description which is not accurate. If as old as stated, they were not made of mahogany, nor could they in the widest sense be considered 'Chippendale.'

Somewhat nearer the mark was the man who had picked up somewhere a set of chairs, and described them as perfect 'shield-back *Chippendale* chairs.' They might have been either 'Chippendale,' or with shield-shaped backs, but they could hardly have been both, and if he had described them as Heppelwhite chairs would probably have been correct.

Such instances of popular misconception about furniture might be multiplied almost indefinitely, and it is of course impossible, in a limited space, to do much more than hint how the reader may, if he will, form a tolerably correct idea about English furniture. Even those who might be supposed to know something about its history, cabinet-makers themselves, have usually devoted little attention to the subject, so that it is hoped this short attempt to trace the development of furniture may not be without interest to them as well as to the general reader.

To deal with the furniture of the older civilisations, or even with that of other European countries, is, of course, out of the question here. The remarks must be taken solely as referring to that of England, or, if the reader prefers it, of Great Britain, and more as having a general bearing on that of the present day than of being an attempt to give a detailed history of that of bygone years.

To all intents and purposes, there was little or no furniture made or required prior to Tudor times for the great bulk of the population. Of course, no doubt people had seats and tables, or substitutes for what we should consider such, but domestic comfort was not studied to any great extent. Even the higher grades of society had very limited ideas on this point, and their furniture seems to have consisted chiefly of chests. To the manufacture of these some degree of attention seems to have been given, but the woodwork otherwise was principally in the form of fixtures, that is, it formed part of the building, or else it was of a more or less temporary character, tables, for instance, being often no more than loose boards laid on movable trestles. That there were exceptions is not denied, but the bulk of the furniture, even if, according to our present meaning of the word, it existed at all, was of the rudest and crudest description. Earthen floors strewn with rushes did not conduce towards refinement, and the general conditions of life were such as to prevent anything beyond coarse, rough furniture of the simplest kind being required. The builder, the joiner, and, when decoration was wanted, the carver were the important wood-working craftsmen, for the cabinet-maker was non-existent; nor is it clear when his business originated as a separate one. That wood-working was a craft which had made considerable advancement, but not in the direction of furniture making, is undoubted, and as the years roll on we find that comparatively small wooden construction received more attention. As civilisation increased so did the need for greater domestic comfort manifest itself, with the result that articles of furniture became more numerous and convenient. The history of a people may almost be said to be written in their furniture. It shows clearly their progress, altering with their habits and customs, adapted to them in fact. Leaving the past for a moment for the present, we recognise this, not only in the differences which are found in different countries, but in the adaptations to peculiar needs. Thus, though the same style may more or less prevail in both, the characteristics of the furniture of a ship's cabin and of a dwelling house are very decided. So has it always been.

Not till towards the end of the Tudor period or the commencement of the Stuart dynasty can it be considered that any marked advance had been made in furniture, so that for all practical purposes its history may be said to date from the latter half of the sixteenth century. At that time, during the reign of Elizabeth, considerable attention was evidently paid to the construction and adornment of movable wooden constructions, *i.e.*, articles of furniture, though according to modern notions these were anything but comfortable or convenient.

It is from this and slightly subsequent periods that much of the spurious antique oak furniture so frequently met with purports to belong, if one may judge from the carving which has been so liberally bestowed upon it. Although much of the furniture, such as it was, and principally consisting of chests, chairs, settles or long seats to accommodate more than one person at a time, with here and there a cabinet or wardrobe or a table, was no doubt carved, it must not be forgotten that most of it was comparatively plain. The furniture was more for use than for ornament, and only in the choicest work belonging to the wealthy could much carving be indulged in. Somehow or other the notion seems to be prevalent that all or nearly all of the furniture of the period referred to was carved. It seems very unreasonable to suppose that it was so, for otherwise one is at a loss to account for the existence of so much plain or very slightly carved pieces of old oak furniture which are to be met with in almost any part of the country. True, in the absence of dates on them, it is difficult, impossible indeed, to say exactly when such things were made, though there are often sufficient indications to indicate that they are seventeenth century work, and that, therefore, they ought in the opinion of many to be carved in characteristic style. If any one doubts the existence of so much plain or uncarved oak work, let him just note what goes *in* to any considerable antique furniture dealer's workshop. He will find that comparatively little of it is carved, and that coffers or chests, settles, chairs, and suchlike things, are in the majority, though among the more modern articles will be bureaus, clock-cases, &c. By-and-by these all make their appearance in the show-rooms or are offered to the purchaser, but no longer plain. No ; they are, to use an expressive though not very elegant phrase, ' smothered with carving.' If the original maker left the things plain, according to the restorer he made a mistake, and it is one which he, the restorer, considers it his duty, or perhaps I should rather say, his interest to rectify.

I am hardly going beyond the mark when I assert that it is almost impossible to obtain a really genuine unspoiled piece of old oak furniture which has had the misfortune to pass through the hands of a dealer or restorer. That some of these may be conscientious in their work I do not deny, but it is a lamentable fact that mostly they do far too much of what can only be called by courtesy restoration and repairing. That old things, when they come into their possession, are often sadly in need of repairs is undoubted, and if the work were limited to doing what is necessary, no objections could be raised ; in fact, the restorer or repairer would occupy an honourable position. When, however, he alters the style of the thing entirely by carving it when the original maker left it plain, or as is not uncommonly done, pulls it altogether to pieces and forms an entirely different article with them and portions of others, the work is not honest. The thing is palmed off as a piece of genuine old work, repaired and cleaned up. Nor is the restorer or 'faker' of old work content with structural alterations merely, for either from his own ignorance or from a desire to pander to that of his customers, he does not stop short at simply cleaning the wood from dirt, but darkens it and varnishes it till both the colour and figure or grain of the wood are hidden.

Oak of a moderate age is not black, but a brown more or less deep, and altogether different from the colour which the restorer gives it. The blackness, when it is not some stain purposely put on, simply results from dirt and smoke, and should rather be washed off than be added to. By means such as these innumerable articles of old furniture are utterly ruined every year by the vandals who deal in 'antique.' I do not now refer to the fabrication of sham antique furniture in its entirety, that is entirely new articles made and sold as old things, an industry unfortunately of considerable extent, so much as to the alterations which are made in really old work. The things themselves are old, so that the dealer may be correct in selling them as such; but, unfortunately, he does not think it necessary to explain that much of the work is entirely modern. As many of those who devote themselves to this kind of work are exceedingly skilful in imitating common rough old carving, it is not always an easy matter to distinguish between the genuine and the false, though there are usually sufficient indications to guide those who have studied the subject in arriving at a correct opinion. It may be easy enough comparatively for the student of old furniture to do so now while the work is fresh, but as it gets assimilated to the genuine the difficulties of distinguishing between the two will be greatly increased. In time it will be impossible to decide which is false and which is real.

Wardour Street enjoys a reputation of a kind for antique furniture as well as for other articles which are sought for by the curio-hunter. Even in it, now and then, a piece of genuine unsophisticated old oak furniture may be met with, but even when the basis of its construction is what it purports to be, it too often happens that the whole article has been tampered with to an unwarrantable degree. Those who wish to study old furniture need not think of being able to do so in any dealer's establishment unless in the workrooms, a part of the premises not usually shown. But perhaps the novice may think that in country districts he may be able to meet with genuine carved oak-work even in the hands of dealers. He may be. In my experience, however, and it is an extensive one, it is seldom possible to buy any piece of antique oak furniture which has been restored for purposes of sale, with any expectation of its being a genuine specimen of old work. The system of falsification is pursued as thoroughly in remote towns and villages of country districts as in London itself. If a piece of old oak furniture is plain there is a chance that not much has been done to it in the way of 'faking it up,' but as soon as he comes across an elaborately carved piece the novice should, at any rate, be careful to ascertain its history. Of course, this can generally be got from the dealer, but after what has been said the probability of its being authentic or otherwise need not be dilated on. Of course, I am not alluding to furniture in antique style made and sold by cabinet-makers of repute, for they would not endeavour to sell it for other than what it is, viz., modern work produced in the Elizabethan or other style. This, of course, is perfectly legitimate, and altogether different from the fabrication of sham antiques. Even dates are not altogether to be relied on as proving the genuineness of the work, for there is no more difficulty in carving them than any other device.

From all the foregoing it will be seen that the study of genuine old furniture, especially of that anterior to the last century, is not altogether an easy matter, and the young cabinet-maker must be specially careful not to be deceived by the sham carving. The best way to study the real is by means of specimens in museums, and in existing parts of old buildings. Many pieces of furniture which purport to be old he can, however, unhesitatingly pronounce to be more or less false, when he finds them got up as 'carved antique oak.' Common sense will often be a sufficient guide. For instance, no one could by any possibility expect to find a hat and umbrella-stand dated back to the reign of Queen Bess, while most of the carved bureaus which are seen can generally be attributed to modern skill. To sum up this part of the subject, the variety of articles of furniture made in this country till well within the last two hundred years was limited, and comparatively little of that used was decorated. Then, as now, the finest work belonged to the well-to-do.

One has almost got into the habit of regarding all the woodwork, the furniture of the period alluded to, as having been oak, and undoubtedly it was the principal timber employed, though other native kinds were used.

By the latter part of the seventeenth century the importance of furniture as a separate business had considerably increased, though there was comparatively little alteration in its general style till after the accession of William of Orange. It may be interesting to note here a sentence from the diary of gossipy old Pepys, under date January 1, 1669. He writes: 'To the cabinet shops to look out, and did agree for a cabinet . . . . and I did buy one, cost 11*l*., which is very pretty, of walnut-tree.' This is noteworthy as showing that cabinet-making was recognised as a separate trade, and that oak was not exclusively used.

During the reigns of William and his successor the style of furniture altered considerably, domestic arts and architecture having received considerable impetus. The construction became lighter, if not more beautiful, though this is a good deal a matter of taste. The style, it must be confessed, was somewhat nondescript, and on it the modern 'Queen Anne' is more or less founded. Although, doubtless, much of what we roughly call Elizabethan style continued to be made, for fashions did not rapidly alter in those days of slow travelling, a distinct advance was made. The old kind of work did well enough for the country joiner or cabinet-maker, but those with any pretensions to fashion wanted the latest thing out—just as they do now. The principal difference now is that the fashion spreads more quickly.

During the reigns of Queen Anne and George I., furniture-making developed into an important industry, fostered, among other things, by the introduction of mahogany, as well as by improved ideas of domestic comfort and luxury. I am inclined, however, to think that the introduction of mahogany gave a stimulus, perhaps, greater than anything else to the designing and production of furniture. It marks an epoch.

The story of its introduction is as follows:—Some pieces were given to a Dr. Gibbons by his brother, a captain in the merchant service. The doctor, who was at the time having a house built, wished the wood to be used in connexion with it, but the joiners declined to work it, on the plea that it was too hard for their tools. Ultimately a cabinet-maker named Wollaston took it in hand, and made a candle-box, a piece of furniture now obsolete in good houses, from some of it. Apparently, the beauty of this box was much admired, for, as an earlier writer on the subject says, 'it outshone all the other furniture of the doctor, who gave Wollaston the remainder' of the wood. From this Wollaston made two bureaus, one of which belonged to Dr. Gibbons, and the other to the Duchess of Buckingham. From this time (*circa* 1724) the use of mahogany rapidly spread, though if the dates attributed to more than one well-authenticated piece of furniture are to be relied on, mahogany must have been used occasionally from an earlier date. Thus, the Worshipful Companies of Iron-mongers and Carpenters each possess chairs said to have been made *about* 1700.

Naturally the construction of mahogany furniture led to alterations in style and methods of working it, till towards the middle of the eighteenth century we find the style now generally recognised as Chippendale had gradually developed, though hardly at its height till a few years later. It may be explained that Chippendale was a cabinet-maker, a man of undoubted ability as a designer, and of considerable position in his trade. As maker to the Court, it may at once be conceded that his work was fashionable and of good quality, both as regards workmanship and materials. That the style generally associated with his name was originated by him is, however, more than I for one can think. He undoubtedly did much to popularise it, even to improve on the lines of his contemporaries and immediate fore-runners, but probably nothing more. He, as one of the leading cabinet-makers of his day, no doubt had an influence, but for the rest he probably followed the prevailing fashion. The prominence given to his name now is chiefly owing to his book of designs, first published in 1754. In it, of course, we get the purest 'Chippendale,' and if the designs were all made by himself we cannot but admire his versatility, although the style is not all that could be desired, indeed mostly quite the reverse. It is rather a conglomeration of styles than anything else, for we find Gothic, Chinese, or what was supposed to be such, rococo, &c., in strange medley. As a rule, the simplest of his designs are the best, and in the later editions of his work there can be little doubt that he was assisted in his productions by others.

Strictly, the term Chippendale could only be applied to furniture made by him, or perhaps it should rather be said under his superintendence, or from his own designs, but it is commonly used in connexion with any furniture of similar style. To trace the causes which led to the development of the so-called Chippendale style would be an interesting task, but unfortunately space forbids, and it must suffice to state that those who wish to study it cannot do better than consult his book of designs. It is somewhat rare, but can be seen in the South Kensington and other important art libraries. Another book of designs, Manwaring's, published about 1766, shows furniture of precisely the same style, but more florid and without the refinement of Chippendale's. There are also several other books of the same kind, all proving that whether Chippendale originated the style known by his name or not, he certainly did not enjoy a monopoly of it. The work was principally in mahogany, and is noted for its careful finish; but it is extremely difficult, if not altogether impossible, to say with certainty whether much that is called by his name was made in his workshops or in those of his contemporaries. Naturally these combined made more than he did.

Following Chippendale, the next great cabinet-maker who illustrated a distinct style was Heppelwhite, whose designs are of a simpler character, for a reaction evidently soon set in against the florid monstrosities of Chippendale and his contemporaries. Much of the furniture called Chippendale was made during the Heppelwhite period; not one of long duration however, for not many years after the publication of his designs we find one by Sheraton, who, from an artistic point of view, may be considered the greatest of the three masters who did so much for furniture in England during the last half of the eighteenth century. His designs, that is to say his earlier ones, show a keener appreciation of art principles than those of either of his predecessors, while his later ones show a distinct falling off. Whether this may be attributed to his own failing skill or to the influence of fashion exercising its hold over his pencil it would be useless to discuss, for there is a good deal to be urged in favour of both arguments. In addition to cabinet-makers there were many architects contemporary with Chippendale, Heppelwhite, or Sheraton, who designed furniture or were considerably interested in it, among others the Adam Brothers. An examination of the work of any of them shows it to have been more or less in harmony with that of those who were exclusively cabinet-makers, and it will therefore be unnecessary to refer to it more specifically.

The retrogressive movement in cabinet-making as an art, already begun in Sheraton's time, say at the very commencement of the present century, continued with little abatement till well within the memory of those who would feel hurt to be called old men. A dead level of hideousness prevailed without a single redeeming feature to relieve it for at least the next fifty years, unless the undoubtedly good work which was thrown away in executing bad designs can be considered as such. I do not flatter myself that many old cabinet-makers, that is, men who were in their prime as artisans even so late as thirty years ago, will read these lines; if they do they will probably not agree with these latter remarks, for strange as it may seem there are still to be found among them many who consider the now old-fashioned loo table with massive pillar, carved claws, and figured top, with other furniture in the same style, as nearly perfection as possible. Such furniture may have possessed some beauty, but it lay entirely in the wood, and by no means in the design, which was mostly cumbersome and heavy, devoid of taste, and too often ill adapted to its purpose. Apart from the manual dexterity evinced, there is little worthy of admiration in furniture made since 1800 till comparatively recently. The improvement which has been noticeable in the last few years has been attributed to the influence, or as having originated from the Exhibition of 1851, though for many years after there was little appreciable difference in the quality of the design. Even so lately as twenty years ago the writer remembers a discussion among the heads of a leading firm in the furnishing trade in London on the subject of what was then only slightly and somewhat disparagingly known as 'art' furniture to distinguish it from the other kind, which certainly was not artistic. The decision arrived at was that though art furniture might do for certain high art advocates, it would remain caviare to the multitude. At that time the furnishing establishments who could supply really well-designed furniture were few. What do we find now? Why, that every insignificant furniture dealer professes to be able to supply art furniture and fit up houses in an artistic style. The man who admires the elegant drawing-room suite of five-and-twenty years ago, with the couch or settee on which no one could recline with comfort, and the accompanying chairs with protuberant carvings just placed so that they would make their presence felt on the sitter's spine, all of them covered with striped rep of gaudy hue, is behind the age. Whether the purveyor of modern art furniture knows anything about the principles of art as applied to wood and furniture is another matter altogether. Still, furniture is better designed than it used to be, and the dealer must necessarily supply what is fashionable, so that it is not difficult in most places to meet with well-designed furniture, if, from causes which were alluded to before, it is not always so well made as it might be.

To what, then, may the change which has come o'er the spirit of the dream be attributed? I am inclined to attribute the improvement to the general advancement which has taken place, not only among those interested in the production of furniture, but to the increased attention which is paid to art throughout the entire community. In justice also it may be said that much is owing to that useful, though, outside the trade, little-recognised body of men, the professional designers of furniture. Many of them are not only thoroughly trained artists in wood, but have a knowledge of the conditions under which furniture is made, and must, therefore, be distinguished from those amateur designers who, however keen their general appreciation of art, have not that special knowledge of furniture without which no man can successfully design it. I say this because many such people have arrogated to themselves an importance as teachers of taste and art applied to furniture which is by no means warranted. They may be more or less acquainted with the broad principles of art, but they are quite unable to adapt it to furniture. Even architects, who are often supposed to be competent to design furniture, are, unless they have made it a special study, lamentable failures. When, like the late Bruce J. Talbot—a name almost unknown outside the furniture trade—they have taken the trouble to understand furniture, they can design it as well as those who have made it their special avocation. Otherwise, there is not a manufacturer who has had to work to the designs of an ordinary architect who could not tell some funny tales about them. They do not seem able to get away from the idea that they are designing buildings instead of contents of buildings. As mere drawings their designs may look very nice, but when critically examined the mistakes to a cabinet-maker are often ludicrous. There is a common idea that furniture designed by architects is not only more artistic, but more substantial —better in every way than any other. This is a great mistake, but if there is any difference in quality it is solely owing to the fact that when an architect is employed the price admits of good work being done. Men like Timms have, in the exercise of their profession of furniture designers, done more to popularise carefully-made, well-devised furniture, unknown though their names may be to the general public, than all the art teachers, architects, and others of the amateur class put together. They, the general practitioners of art, have little weight in the practical everyday life of the workshop, for they are not in touch with the workers.

It may seem somewhat contradictory to these remarks to attribute any important share in the upward movement to the late Sir Charles Eastlake, but there is no doubt that his book, modestly called *Hints on Household Taste*, had much to do with it. As an exposition of art-principles applied to furniture, I know of nothing superior to it, and those readers who wish to design furniture for themselves cannot do better than study it. I must, however, caution them that all the arguments, plausible though they may seem, must, when they concern practical matters, be accepted with the greatest reserve. This is notably the case in connexion with veneering, for there is much that is incorrect when regarded from a practical point of view, and altogether apart from the broad principle that this mode of finishing is open to abuse. Notwithstanding these blemishes, and they are commonly met with in all books, however otherwise valuable, written by those who have little or no acquaintance with the practical side of furniture, with its actual manufacture, the young cabinet-maker may be recommended to read the book referred to. In it will be found the germ of the so-called 'Eastlake' style of American furniture, though much of it is only so in name.

Our own 'art' furniture of twenty years or so ago, though with certain modifications it has remained in favour ever since, was more truly 'Eastlake.' The distinguishing features may be said to be plainness and severity of line with solidity of construction—a solidity often more apparent than real. Of course, when carried out in its integrity, furniture in the 'Eastlake,' or, as it was more generally though erroneously called in this country, the 'Early English' style, was, and is, soundly constructed, but there is no greater difficulty in making cheap common work in it than in any other. Our art-furniture critics often seem to forget this, or to be unaware that the solid, severe-looking articles of furniture, call them Early English, Jacobean, Queen Anne, or anything else, may be, and often are, as flimsy and deceptive in construction as the greatest rubbish of pine and 'knife-cut' veneer put together to imitate the lordly Spanish for sale 'on the hawk' in the Curtain Road.

The 'Early English' furniture, or, more correctly, furniture in Old English styles, of recent years—there was, as I have endeavoured to show, none to speak of in the times which its name would indicate—was plain and severe. Such decoration as there was consisted of beads, small mouldings, and chamfered edges with a little carving, distinguishing features which are still much in vogue. Veneered work seemed a thing of the past, unless for small fancy articles and common things. The taste of the times was altogether against the large veneered carcase work of mahogany. For almost the first time since its introduction this wood ceased to be popular; it was no longer *the* wood for good furniture. In place of it ash and American walnut were used in bedrooms, oak, often fumigated and wax-polished, for dining-rooms, and black, or, as it is generally called, ebonised furniture for drawing-rooms.

It is interesting to look back and note the changes which have occurred even within a few years, not so much in the general construction of furniture, as in the decorative details. First, there was the rage for gilded incisions and panels, the latter with painted ornament-ations, which, when well done, often added greatly to the beauty of the furniture. The black and gold painted panels did not, however, enjoy a long run of popular favour for the best furniture. The paintings became worse and worse, till, instead of adorning, they spoilt the effect of furniture which was perhaps otherwise well designed. Black and gold decoration got common in fact, and therefore became, or was considered, unsuit-able for good things. This is also the reason, or one of the principal ones, for the rapid changes which have taken place in recent years.

A style or feature of decoration has scarcely been well introduced in furniture of the best class than it is repeated in the commonest work without discretion. It gets overdone. Thus at one time turned spindles were all the rage; hardly a piece of furniture was made without one or more rows of them. The turners had a good time then in the wholesale furniture-making districts. Turned spindles are all very well in their way, but one does not care to see them here, there, and everywhere. People got tired of them, so they went. Then the places which were previously fitted with them were filled with fretwork, and the fret-cutters of Bethnal Green had their innings. As time went on, ebonised work palled. American walnut ousted it from our drawing-room so completely that it very rapidly became old-fashioned. The same kind of walnut was also more extensively used both for bedroom and dining-room, and, of course, for other furniture, till the consumption was such, that from having been an almost valueless timber, it was in considerable demand, and has remained so since, though apparently it is no longer used so much as formerly, for it, in its turn, has been displaced to a considerable extent. Rosewood, for a long time almost forgotten, except as an old-fashioned wood, gradually came to the fore, mostly inlaid with marquetry adapted from the style of the Italian Renaissance. The marquetry-cutters then became the masters of the situation, and even very young cabinet-makers may remember what a difficulty there was in getting the inlays, and how often a job was kept waiting for them. Of course, that soon righted itself, only at the same time marquetry, which at first was only seen in good work, was gradually reduced in quality to correspond with much of the furniture in which it is now found. Along with rosewood, which is much used for drawing-room furniture, mahogany, mostly darkened in colour by staining or fumigating, has been more extensively used again.

The extreme severity of line apparent a few years ago has also been considerably modified, and running from one extreme to the other, it almost seems that in the immediate future we may expect a return to the curves of the Chippendale period, though doubtless somewhat less exuberant.

Such is a brief and only a very curtailed outline of the changes which have taken place within the last five-and-twenty years, for it has been impossible to do more than glance at some of the principal features of the furniture made during that period. It will, however, serve to show the young cabinet-maker that changes are constantly going on, and that to adapt himself to them it is necessary to be able to do more than one kind of work. Thus it would be little use for him to be able to do only solid work, and to know nothing about laying veneers, although these for a time might seem to have gone out of use. As the outline of the development of furniture has been necessarily of a very sketchy nature, it may be advisable to warn the student that there is no such thing as an abrupt transition from one style to another; the changes are so gradual that while they are going on around us their progress is hardly noticed. By degrees we awake to the fact that the fashion has undergone a change, but exactly how or when who can tell? It is sometimes urged as a reproach that we moderns invent no new style; that all we can do is to modify those which have existed, even, as some say, to copy them. They tell us we are reverting to the work of the old masters of the craft and copying it. That now and then a piece of furniture may be copied is not to be gainsaid, but the furniture-designer must do far more than this, for there is very little indeed of any old style which is copyable for ordinary purposes. All that can be done is to seize the leading features of any given style, and incorporate them in a design adapted to modern requirements. This is an altogether different matter from mere copying, as any novice who tries will soon find out.

The beginner is advised not to try and invent something quite new in style, but rather to acquire a good knowledge of what has gone before. He will then have comparatively little difficulty in producing sufficient novelty. The remark of Sir Joshua Reynolds is well worthy of note, that 'Invention, strictly speaking, is little more than a new combination of those images which have been previously gathered and deposited in the memory.'

It is the custom in some quarters and by some writers to urge that furniture is not made so well now as in times gone by, and were this the case it would certainly be a slur on the capability of the modern cabinet-maker. Over and over again antiquated forms of joints and methods of construction have been advocated in preference to those usually adopted. The reasons given for such advocacy are apparently based on an entire misconception of modern work of the better kind, with which alone we have to do. Thus one well-known writer states as a remarkable fact that he was asked more for a piece of furniture made of solid wood than for the same pattern veneered, and argues from this that veneering is bad instead of being merely a question of price. Another writer (Eastlake), in the book already referred to, notes that he had a substantial oak table made for less money than one veneered with walnut or rosewood would have cost. This, of course, is only what any cabinet-maker would expect if the veneered table was to be made of equal quality with the solid one, for the cost of the veneers and of the extra labour must be added. Many of those who have not sufficiently considered the subject would have cabinet-makers adopt obsolete methods of construction on the plea that they are so much more durable than those mostly resorted to. As it is desirable that the cabinet-maker should be able to refute such arguments, it will not be amiss to devote some space to their consideration.

The first reason which may be taken up is the alleged superior durability of old furniture, and at the first glance it seems a very good one, though when looked into, there is a good deal to be urged against it. I think we may safely conclude that the specimens which are found remaining were fairly well made; that is to say, substantially put together. The inferior things have been and are being gradually destroyed. The better ones have more care taken with them, and naturally last longer. In other words, it is a case of the survival of the fittest. The rubbish that was made in, let us say, the seventeenth century, has long been destroyed; only the best remains, and it seems but reasonable to suppose that this applies to anything made since. As every one must be aware, it is not the common furniture of the kitchen which has the most care taken with it, but the more costly and highly finished articles. These will remain when the others have been broken up and destroyed, so that it is not at all improbable, when all the poor stuff of to-day has ceased to exist, in time to come the superiority of the furniture of the latter part of the nineteenth century may be proclaimed as forcibly as that of past generations now is. We are frequently told that old work was so much more substantially put together than the modern is that we are rather apt to forget that the weakest of it must necessarily have perished first. A great deal of what remains, even from the last century, is in a very rickety condition, unless it has either had considerable care taken of it or has been restored and repaired since. We look in vain for the overwhelming superiority of old work when it is examined.

Although a great quantity of very poor stuff, rubbish in fact, is made now, it must not be forgotten that there is also a great quantity soundly and honestly constructed, and that it is increasing. By this rather than by the other should the capability of the modern artisan be judged. There is no doubt whatever in the minds of those who have an intimate knowledge of furniture that at no time has better work been done. Of course, if buyers will hunt around for so-called bargains and buy the cheapest furniture they can get, it is not to be expected that they will get hold of really good work. They should not form an opinion on the quality of modern work from experience so acquired; and it may be a matter worth noting that those who prate most about 'good old work' and the paltriness of the modern are not always the most liberal with their tradesmen. They want 'things for next to nothing,' and they get quality accordingly. If people will pay for it they can get furniture quite equal in quality to that made by those who are dead and gone. The skill of the modern cabinet-maker has in no degree departed, but, on the contrary, it has improved. Compare the crude, rough work which was made in the seventeenth century with the neatness and superior finish of that of to-day and note the advances which have been made.

The return to the solid construction which prevailed during the period last mentioned has been advocated by more than one theoretical writer, but before this can be agreed to by the modern cabinet-maker there are one or two points on which he should be clear. The mortised and tenoned and pinned-through joint has often been trotted forward as an example to be copied. Well, the pinned joint was right enough for the times during which it prevailed, and in the absence of the superior kinds now prevalent. When tools were rude, great precision and nicety of finish could not be expected, but the case is different now. To return to the crude joints of our ancestors would be a distinctly retrograde move. The tools which the modern cabinet-maker may use are of the finest quality. Screw nails and good glue are obtainable in every village, good designs may easily be studied, and the intelligence of the artisan certainly has not decreased. In these circumstances it does not seem reasonable to expect a return to the rough style of work of the country carpenter or mediæval joiner, unable mostly to read or write and wholly uneducated, save in the crude customs of his craft. He could not use screw nails, nor the many advantages which are now available, because they did not then exist. I do not think his wooden pins prove that he used them because he considered them superior to anything else, but rather because there was nothing else which he could easily obtain or make. If there is any lesson of a practical kind to be learned from the study of old work, it may be considered to consist principally in the fact that the worker made the most of his opportunities, and that apparently he was not in a hurry. Of course, as far as design goes, there is much to be learned by the study of the woodwork of any period, but I do strongly wish to warn my readers against copying the roughness or crudeness which is found in nearly all English so-called antique furniture. On the contrary, let them—while not neglecting sound construction, without which no furniture can be really beautiful—avail themselves fully of modern tools, appliances, and methods of work.

As regards the strength of furniture, it must not be forgotten that extraordinary substance and massive construction are not required. The things are indoors, and not exposed to the weather, therefore we do not require them as 'strong as a house;' nor are they in well-regulated households exposed to rough usage. The strength, then, will be sufficient if the things can be fairly used for their ostensible purpose; thus it is not necessary to make a small fancy table for the drawing-room so strong as one which is intended to bear heavy weights. If furniture were constantly exposed to the weather or to rough usage, one could understand there might be some reason for returning to the strong carpentry of the sixteenth and seventeenth centuries; but as it is, the cabinet-maker may be quite contented with good modern construction, and he can well afford to ignore the advice of dilettante writers about it. Careful and intelligent work is wanted, not a return to old methods.

The professional cabinet-maker is often hampered in his desire to do good work by financial considerations; the thing must be done within a certain time, or there will be a loss on it. The amateur artisan, however, is not under any such restrictions, so that he at any rate will be able to make his work as thorough as his skill will allow.

Perhaps before concluding this chapter something should be said about the suitability of cabinet-making for amateur work, for it is undoubtedly popular in one form or another. The work is neither very laborious nor very difficult, little more than a healthy exercise for mind and body. To suppose that any amateur would wish to furnish his house entirely with his own work is hardly reasonable, but he may make many things from time to time which will add considerably to its comfort and appearance. If he does not care to make anything of large dimensions, there are plenty of smaller articles which may engage his attention, and afford him much pleasure in the making as well as afterwards. The cost of the necessary tools need not be great, and if bought with discretion it will not be long before they have paid for themselves in the value of the work done by their aid. In the following pages it is to be hoped the most unskilled will find all necessary information for making furniture.

CHAPTER III.

FURNITURE WOODS.

Mahogany — Cedar — Pencil Cedar — Oak — Walnut — Ash — Hungarian Ash — Rosewood — Birch — Beech — Satinwood — Pine — Pitch Pine — American Whitewood — Sequoia — Timbers occasionally used — Logs — Buying Timber — Measurements — Seasoning and Drying — Levelling Boards — Waste.

IN the construction of furniture, wood is of course the chief material, though there are many others, such as glue, nails, &c. These will be found described in due course.

Wood, from its importance, naturally occupies the first position, so that some space may be devoted to such information about it as it is necessary for the cabinet-maker to possess. Some of the principal woods used have already been incidentally referred to, but a more minute account of them, as well as others which are largely employed, may now be given.

Mahogany.—With this, as the one which is in popular ideas most intimately associated with furniture, a start may be made. To describe its general appearance is of course unnecessary, as it is so well known. Apart from its beauty, it has several features which render it peculiarly suitable for furniture. It is a wood which stands well, that is, it is not apt to twist and split after it has been seasoned and worked up. It is also clean— a word which, it may be explained, is not to be taken in its ordinary signification when applied to timber, but rather as meaning that the wood is free from knots and other defective markings. Mahogany, it may be observed, is remarkably free from knots, and is, therefore, apart from the hardness of some varieties, workable with comparative ease, and is susceptible of a high degree of finish. In some instances the boards are of considerable width, so wide that almost the widest pieces, such as sideboard tops, wardrobe ends, and so on, which the cabinet-maker uses, may be got without joining one or more boards together to obtain the necessary width. This, of course, is often an advantage. There is probably no wood in which so many varieties are found and having such a wide range in value. From plain wood, with no figure to speak of and little more costly than the best pine to choicely marked Spanish, there is a great difference, for the latter is one of the most costly of timbers. When very finely marked, it is too valuable to be used in the solid; *i.e.*, it is cut into veneers, and some of these are worth high prices.

The cheapest and commonest mahogany is the Honduras, or, as it is often simply called, Baywood. This has little figure, at least the typical Baywood, for varieties are met with which have a fair marking, is very clean, and from its comparative softness is easily workable. It is admirably adapted for any purpose where figure is not essential or where plainness is not objectionable. As a groundwork for veneering on it is unsurpassed, especially when the veneer is a choicer mahogany. It is also admirable for ebonising, as it takes the stain and finishes well. Compared with the finer varieties of mahogany, the grain is coarse and open.

The best mahogany is that known as Spanish, a term which may be somewhat misleading to the novice as implying that the wood comes from Spain. It is, however, a West Indian production. Strictly speaking, Spanish mahogany is that from San Domingo, but for all practical purposes the exact place of growth or port of shipment is of small consequence, and the definition is by no means closely adhered to. Instead of concerning himself with the nativity of the wood or the place whence it was shipped, it is better for the cabinet-maker to make his selection when purchasing according to the figure and general quality of the wood. Although Honduras and Spanish have been referred to as the plainest and best kinds of mahogany, it does not follow that all of the former is figureless or that the latter is always well marked. On the contrary, Honduras is often found with a considerable amount of figure, though seldom of the finest, while the plainer kinds of Spanish may not be a bit better than it. The novice is therefore cautioned not to put too much importance on the fact of a piece of mahogany being called Spanish, but to use his own judgment whether any particular lot, whatever it may be called, will suit his purpose. The quality of mahogany, or rather its value, is determined chiefly by the amount of figure, though in unusually wide pieces the width must be taken into account too. It may be remarked that no wood improves so much with age as mahogany, which is not seen at its best when new, as the colour becomes much richer in time.

Cedar of the coarse kind, much used for making cigar-boxes, must not be mistaken for mahogany, the plainer kinds of which it very much resembles. It is, however, very useful for drawer bottoms and such like interior work.

Cedar of the fragrant variety, or, as it is generally called, pencil cedar, is an altogether different wood, being fine and close grained. It is very little used in furniture, and then only for fittings in fancy articles like davenports and other small writing-desks. It is soft and pleasant to work, but splits very easily.

Oak, like mahogany, is too well known to require any minute description. There are many varieties, and the choicest of these is the brown or English, as it is often called, to distinguish it from the lighter or Riga and Dantzic wood. Brown oak is very hard, and not an easy wood for novices to use. It is often very choicely marked, and the finest, or Pollard oak, is much used for veneers; in fact, the true pollard can only be made up in this form. The lighter oaks are frequently stained in imitation of brown, to which in colour a very near approach can easily be made. As in the case of mahogany, it is little use for the cabinet-maker to interest himself about the origin of the foreign oak. I have seen some excellent 'Dantzic' which came from South Germany, and it was none the worse on that account. There is a good deal of American oak used now. It is rather different from the European. Much of it is perfectly plain, and some of it of a peculiar pinkish tinge. This latter should not be used in any article intended to be darkened by fumigation with ammonia. When American oak is of good colour and figure, it is quite as suitable for furniture as any other kind is. The perfectly plain oak is often called and sold as coffin wood, and being cheap, is sometimes useful for inconspicuous parts of furniture. The figure of light oak, it may be remarked, consists entirely in the hard, shiny looking marks, which form such an important feature in it that they cannot be mistaken. Though hard, oak is by no means a difficult wood to work. It should be very dry and well seasoned before it is made up.

Walnut.—The kind now chiefly used is the American or black walnut. It is a hard brown wood with very little figure, generally very clean, and often of considerable width. From having been regarded as a worthless timber in this country it has now become one of the most valued for furniture, and commands a price which seems altogether beyond its beauty. Other varieties of walnut are more ornamental in appearance, the popular name applied to them being Italian, wherever they were grown. They are comparatively little used now, as apart from figure the American is a better timber for cabinet-making purposes. Much of the veneer which is seen is of the variety known as 'Burr' walnut, and is very beautifully figured. It can only be used in the form of veneers. The wood known as satin walnut must not be confounded with any other, for it is a distinct kind, and the name is more or less a fancy one. It is of a yellowish or light brown colour, with dark markings, and cannot be recommended for good work, as it is unreliable, though often used for bedroom furniture. Till some one dubbed it satin walnut the timber was not in repute.

Ash is very popular for bedroom furniture, little else being made of it, though, beyond custom, there is apparently no reason why it should not be worked up for other rooms. As is no doubt well known, it is light in colour, tough and hard, with a moderate degree of resemblance to oak. As a rule it has little figure, and what there is is coarse. The American is the best for furniture, as it is generally of better colour than the English, and the beauty of ash is considered to consist mainly in the lightness of its colour.

Hungarian Ash is a totally different kind of wood, seldom, if ever, used except as veneers, in which form it is sometimes seen. It is finely marked, but is by no means indispensable to the cabinet-maker, as it can hardly be classed among the popular woods.

Rosewood.—This is a handsome timber, which, after having been out of the fashion for many years, is once more considerably used, especially for drawing-room furniture. The colour is a dark red or brown, with strong markings of a much deeper tint, frequently almost black. Like mahogany, the colour deepens considerably by exposure. The chief peculiarity of this wood is its remarkable fragrance, though of late years much has been imported which has little or no smell. This 'bastard' rosewood is greatly used on cheap furniture, and is more frequently met with than the other, which is comparatively costly. So far as mere appearance is concerned, there is not much difference between the various kinds.

Birch, especially American, is a good wood, though not much used, and then chiefly for bedroom furniture. It is much more beautiful than ash, but at the moment is not so fashionable. It will, however, no doubt come in again. It is light in colour, close grained, and often very finely figured. Except in colour, it has a strong resemblance to mahogany, *i.e.*, in the general style of the figuring, and may easily be stained to an excellent imitation of that wood.

Beech is somewhat similar in colour, but not much used except for chairs. It has a close, fine grain, with small figure, and as it can easily be ebonised or stained to imitate mahogany or rosewood, it is sometimes useful.

Satinwood is decidedly a fancy wood, and was at one time much more extensively used than at present. It is of a pale yellow colour, and often very beautifully figured. In fact, it is perhaps the most beautiful and delicate wood which is used. The markings are lustrous, and not unlike those of good Spanish mahogany. Needless to say, the best of it is cut into veneers, in which form it is generally used, and is frequently of considerable value. It is hard and close grained, and consequently susceptible of a very high degree of finish. A pleasant fragrance emanates from it, distinctly perceptible, though not so strong as that of rosewood. It is seldom used except for costly furniture, unless on small fancy articles.

Pine.—This is one of the most useful kinds of timber to the cabinet-maker, either for making up furniture entirely, or for those portions which are not seen, and are of comparative unimportance, such as the backs of looking-glasses and of carcase work generally. It is rare indeed to find a piece of furniture which is made entirely without it, for by using it judiciously the cost of production is considerably reduced without any detriment to the article. Although any kind of pine may be used, all are not equally suitable for furniture, as some of them are knotty and resinous. The most pleasant to work and the cleanest is the yellow pine, which is easily obtainable. When of really good quality, it is very clean and free from knots. The commoner kinds, such as spruce, which do very well for building purposes, cannot be recommended to the cabinet-maker, who in this timber should use only the best he can get. It will cost a little more in the first place, but owing to the small amount of waste, can be used more economically than common stuff. It is also much more pleasant to work with sound, clean wood than with that which is coarse and full of knots. Pine, it may be observed, is often loosely spoken of as deal, though strictly this should be limited to a certain sized piece. Any kind of pine, provided it is free from knots, may be used in furniture, but the sort recommended will generally be found the best. Pine, it is well known, is a soft, easy wood to work, and it is cheap.

Pitch pine, though similar in name, is a different kind of wood, and one not altogether in favour with cabinet-makers, by whom, however, it is sometimes used. In appearance it is darker than the ordinary pine. It has strongly marked figuring often of a very handsome character. It is fairly hard and very resinous, which makes it a somewhat unpleasant wood to work. In addition to this, it is rather unreliable, that is, one apt to twist or split, even though it seems dry and well seasoned.

American Whitewood.—This is an extremely useful substitute for ordinary pine, and has come much into vogue within the last few years. It is somewhat harder than pine, is of a very uniform texture, and of a lightish yellow colour, sometimes nearly white. It is often obtainable in great widths, is remarkably clean, and free from knots, as well as reliable and sound. Apart from its appearance, it has every attribute that can render a timber valuable as a furniture wood. It is generally sold at very low prices, being often obtainable for lower figures than prime pine. It ebonises well, and when suitably stained bears a very close resemblance to American walnut. Owing to its plainness, it does not make such a good imitation of mahogany. It may be used without hesitation either for making furniture of a cheap kind entirely, or for the secondary portions of better articles.

Sequoia, or Californian red pine, has also been much used of late years for inside work instead of pine, and is specially applicable to drawers, where the appearance of a white wood might be objected to. It is reddish in colour, and slightly resembles pencil cedar, being fine and silky, and, like it, splits easily. For this reason, as well as its remarkable softness, it is not so useful as it otherwise would be, and it may be well to note that it is quite unsuitable for general construction. It is very clean, and often runs to enormous widths. It is probably the softest wood known, too soft and spongy to be altogether pleasant to work.

Many other kinds of timber might be mentioned, but to do so would serve no good purpose, as those already enumerated are the principal used in cabinet-making. Speaking generally, every kind of timber may be used, but in practice very few are, and others which have not been specified are of such comparative unimportance that to find them in furniture is quite exceptional. There are certainly some which are used only as veneers, but they will be found mentioned elsewhere. Of course, I do not wish any reader to understand that no others than those named can be used for furniture; on the contrary, there are very many which are equally suitable, only they are either not obtainable regularly, or, what is much the same thing, their advantages have not been sufficiently recognised to lead to their general adoption. Thus there are the Kauri pine and other woods of New Zealand, many of which are admirably adapted for furniture, but are only occasionally seen in this country. Therefore, if the reader finds any timber which he likes the look of, and it seems suitable, there is no reason why he should not use it.

In practical handbooks it is often customary to say a good deal about the shipment and export marks or brands whereby the merchant can recognise certain kinds, and the reader may expect that something of the kind should be given here. I may as well explain that this is not a treatise on the timber trade, and that the cabinet-maker need not concern himself about such details, nor with the measurement of timber in the log. These concern the timber merchant and large consumer principally, so that the space at my disposal may be more appropriately occupied with such information as is likely to be useful to those for whom this book is primarily intended. What affects them as cabinet-makers will be found mentioned, it is hoped, with sufficient explicitness. For those who can use it in sufficient quantities, timber bought in the log will often be the cheapest, but the saving, unless there is considerable experience on the buyer's part, will be inconsiderable, and involve a good deal of trouble in getting the stuff cut up.

I may explain that 'stuff' is the conventional or workshop word which is used when speaking of wood in a general sense, and though it may not be very elegant it is thoroughly well understood among cabinetmakers, so that there can be no objection to using it here. The buyer of logs must have them cut up at the sawmills, have the timber properly stacked, and then wait till it is ready for use, which will probably not be till many months have elapsed. All this involves a considerable amount of labour and loss of time, so that the small consumer will find it, on the whole, more to his advantage to get his stuff in usable quantity and workable condition as he wants it, even though he may have to pay a higher price. In every town of any size timber merchants or dealers will be found who sell the timber in convenient sizes, and in such small quantities that the small consumer need not lay in a big stock. Of course, he must expect to pay rather more than the large buyer, but the difference will not amount to much.

It may be expected—I know it is by some amateurs—that some 'wrinkles' should be given by which the novice may pose as an experienced buyer, and so obtain what he wants at the lowest or wholesale prices. Well, all I can say to those who think thus is that it cannot be done. Want of knowledge cannot be concealed from those who know their business, whether this is selling timber or anything else. The best way to avoid imposition is to buy from a respectable dealer, and by paying fair prices the purchaser will have no difficulty in getting good stuff. If he wants only the lowest priced, then he must not be disappointed if it does not turn out as well as he would like. The small consumer will, in the long run, gain nothing by buying mixed lots at a low figure, for there is sure to be a good deal of waste. The large consumer may occasionally find it to his advantage to do so, for what would be waste to the other may come in for odds and ends.

When buying from a timber-yard, it is seldom that a piece will be cut of any special length that may be wanted. If it is the buyer must be prepared to pay

E

a considerably increased rate. Of course, retailers who lay themselves out to supply amateurs will do this, but their rates all round will be found comparatively high, though perhaps reasonable enough under the circumstances, for it must not be forgotten that to cut to given lengths means inevitably a quantity of unsaleable short pieces left on the dealer's hands. For these he must be compensated, and it will be found, in practice, better to buy a length, even though it may be more than required for immediate use, than to get just what may be wanted. Of course, if the timber is an exceptionally valuable one the case may be different, but the purchaser may be safely left to form his own judgment when this happens.

In large towns, where there is a choice of dealers, it may be well to know that those yards where builders' timber is principally sold are not the best for furniture woods. Some dealers, in fact perhaps the majority, sell all kinds, but in London, Liverpool, Glasgow, Birmingham, and other large centres, there are plenty of dealers who make a speciality of furniture woods and sell it in small or retail quantities. If to be cut into particular thicknesses not in stock, they will undertake the sawing at very moderate prices, selling the piece selected, and charging a rate per foot for sawing. This will save the cabinet-maker a considerable amount of hard labour, which it would not pay him to incur, as the sawing is done at the mills by steam power. Those who live in country places where the better kinds of furniture wood, mahogany, walnut, &c., are not obtainable on the spot, can have it sent from one of the larger centres at very moderate rates for carriage, as the conveyance of timber is not costly. Pine and other common stuff can be got almost anywhere from builders' yards, though when he can do so it will be better for the user to deal direct with a timber merchant. The conditions of trade have so much altered of late years with the increase of transit facilities

FURNITURE WOODS.

and the establishment of sawmills, that many of the difficulties which formerly stood in the way of the small consumer have been considerably reduced.

Although there are many intricacies involved in the measurement of timber, the cabinet-maker has little to do with any of them. The stuff he uses, and in such quantities as he is likely to buy it, is principally quoted for and sold at a rate per foot super. If the thickness is not specified this may generally be understood to mean 1 inch, but as a rule it is stated. Thus, in quotation a price will be given for any wood $\frac{1}{4}$, $\frac{1}{2}$, $\frac{3}{4}$, or 1 inch, as the case may be, so that the purchaser can have no difficulty in knowing the exact cost of material. It should be noted that $\frac{1}{2}$-inch stuff must not be regarded as costing only half the price of 1-inch, for the sawing must be taken into account. For this reason the thinner the stuff the higher its price per foot proportionately to that in the inch. The cabinet-maker seldom needs anything thicker than this, unless in the form of squares for legs, which in places where there is any considerable demand for them are often sold at so much each, per foot run, or per set of four, according to circumstances. When they are not, there is no difficulty in ascertaining their cost either from quotations or by estimating their contents of 1-inch stuff, the price of which will approximately be known.

As the way of calculating superficial measurement may have been forgotten since their schooldays by those whose business does not necessitate its use, the following hints may be of use. The foot super contains 144 ins.; therefore, to get at the superficial contents of any piece of wood, reduce the measurement of length and breadth to inches, multiply them together, and divide by 144. Thus, a piece 6 ft. 3 ins. × 1 ft. 1 in. = 75 ins. × 13 ins., these figures multiplied together give 975 ins., which divided by 144 give 6 ft. and $\frac{111}{144}$ths of a foot, or a little more than 9 ins. Another method of reckoning, and a less cumbersome one, is to proceed as follows :—

```
       Ft.  Ins.
        6   3
        1   1
           ─────
            6  3
        6   3
       ─────────
        6   9  3
```

As boards do not always run of the same width throughout their length, it is usual to take about the average width. Of course, this does not apply to stuff of which the edges have been trimmed.

When the thickness of any wood is specified it must not be expected to measure the full. Thus, any one ordering stuff will find that it is almost invariably thinner than its nominal thickness. This is owing to the waste caused in sawing, for of course the saw removes a perceptible amount in its course instead of splitting through like a wedge. Thus, supposing a plank 3 ins. thick is to be sawn into boards 1 in. thick, there will be the width of two cuts or kerfs made by the saw to be deducted from the thickness of the three boards. As the stuff is only rough when it is sawn, the thickness will be further reduced by the necessary operations of smoothing it down. Thus, nominal 1-inch stuff will probably not measure more than about $\frac{3}{4}$-inch thick down or finished. A proper comprehension of this fact will save many mistakes when setting out work, and it should not be forgotten. The actual difference between nominal and actual thickness varies, but it always exists to an appreciable extent, especially by the time the wood is planed down. Roughly, it may be said to vary from $\frac{1}{8}$ in. to $\frac{1}{4}$ in.

It may be remarked that the quality and amount of figure in wood when rough cannot be easily determined without considerable experience, so that, to the novice, good and bad will look very much alike till part of the surface is cleaned up. It will not do therefore for him to rely much on his own judgment, but rather on that of a respectable dealer. Those who want wood in very small quantities can generally get it from a manufacturer of furniture; and here I would say that many of those who call themselves so do not make anything, but buy everything ready made. It is no use applying to such manufacturers (?) for stuff.

The importance of using only thoroughly seasoned dry stuff cannot be too strongly impressed on the cabinet-maker, and I have heard many amateurs complain that they cannot get wood fit to use. They have bought stuff which has been said to be well-seasoned, and yet when they have used it they have found that it has shrunk, split, or gone wrong in some way or other, leaving them with the fixed idea that they have been imposed upon. Now a few hints may be useful to those who have had such experiences, and prevent others from finding the same fault.

Timber, it must be understood, may be seasoned and yet not be dry enough for immediate use. If it is not dry, it will shrink in becoming so. Moisture or damp causes it to swell, the removal of the damp causes shrinkage, and unless this is unrestrained and even, the wood will probably split. When, therefore, wood fresh from an open timberyard, or from a cold storeroom not artificially heated, is worked up immediately in a warm, dry room, or the article made is kept in one, there is no wonder that it shrinks, or some part of it 'goes.' To those who know anything about it, the wonder would be if it did not. All stuff, even if it is 'seasoned,' should be kept in a warm, dry place for a time before it is made up. The stuff may not be wet or even damp to the touch, probably it is not, but all the same it can hardly be considered dry, from a cabinet-maker's point of view, if it has been stored in a cold place subject to every change of weather.

It is an exceedingly difficult matter to tell when wood is really dry and therefore usable, for whatever precautions are taken to ensure its being so, it really seems sometimes to be of no avail. Every now and then a piece of wood, which to all intents and purposes seemed as seasoned and dry as it could be, will, after months or even years, shrink and split. Few who have had much experience with furniture but could tell strange tales, almost incredible to the general public, about the vagaries of wood. The only thing the cabinet-maker can do is to have it as dry as possible before using it; and it must not be forgotten that wood always contains a certain amount of moisture. It may be only small, but it is there nevertheless, and, practically, it is impossible to get rid of it altogether. If the wood is made absolutely dry by artificial means it will absorb moisture from the atmosphere; hence it is impossible to keep wood perfectly dry.

The amount of moisture contained in what may be regarded as dry, workable wood is perhaps more a matter of scientific than of practical interest, and need not be insisted on if the fact is understood that all timber is influenced more or less by atmospheric changes. In long-continued dry weather wood will shrink, in cold wet weather it will swell. As dwelling-rooms are seldom damp, and are in winter kept warm and dry by fires, it is very easy to see that if furniture is made of damp wood it is sure to split sooner or later when kept in a warm room. All that can be done is to dry the wood well before using it, and this is best done gradually by keeping it for a time in a dry, warm place, similar to that which it will ultimately occupy. Under ordinary circumstances the amateur or small consumer cannot do better than keep it for a time in a warm kitchen. It must, however, not be placed too near a fire, for to do so would probably cause it to twist and split. Of course, the kitchen may not always be a convenient place to dry the wood in, and then some warm, dry place should be selected, more care in this respect naturally being necessary in wet, wintry weather than during a hot summer.

As far as possible the wood should be arranged to allow an equal amount of air all round it. Simply to put a heap of boards neatly on top of each other without any space between would be of little use. If this is the most convenient way to place them, keep them separate by a couple of sticks between each, and see that the bottom board is well raised above the floor. A better way is to stand them on end, always being careful to provide air space between each. If leaning against a wall, especially an outside one, the position of the wood should be changed now and then to ensure the drying being equal on both sides.

Some kinds of timber require more careful treatment than others, and so susceptible are some to atmospheric influences that even after they are, or seem, thoroughly dry, on a fresh surface being exposed to the air by planing, they will cast and twist. Brown oak with much figure in it may be mentioned as a good specimen of this kind of wood.

It is not an easy matter for the novice to tell whether wood is seasoned and dry, and even old hands are occasionally mistaken. The amount of seasoning may by the experienced be fairly estimated by noticing the weather stains and general appearance of the rough surface, but the dryness is somewhat different, and can best be ascertained by actual working. Wet wood is heavy and unpleasant to work, and if the novice will try to saw and plane two pieces of each kind, one purposely wet and the other thoroughly dry, he will easily be able to recognise the difference. I have insisted on the drying of timber more, perhaps, than some experienced cabinet-makers will think necessary, but the novice can hardly take too much care in this respect. Better to have the wood dryer than may be absolutely necessary than the reverse; and I know the tendency when only a small stock is kept is to work up the wood as soon as it is bought.

Very frequently a board after it has been cut for a time does not remain perfectly straight and level. It may twist more or less throughout its entire length or simply become rounded, *i.e.*, hollow on the one side and correspondingly convex on the other. In the former case the only reliable way is to plane it down, and even then it may twist again. Such pieces, therefore, are not the most suitable. Flattening them by weights or pressure is sometimes recommended, but the remedy is little more than temporary.

Boards which are simply rounded may often be brought level either by causing the hollow side to swell or the other one to shrink, though sometimes nothing short of planing down will do. The shrinking of the rounded side may be accomplished by placing the board near a fire, but not so near as to cause it to split, or what is more likely, instead of merely flattening it, reverse the curvature, making the previously hollow side become round. A moderate degree of warmth will soon draw flat a board susceptible to this treatment. The hollow side may be swelled by the application of moisture, and though this theoretically is wrong treatment, when judiciously done harm seldom results. The moisture must, of course, be of the slightest, for to really wet the wood is out of the question. In fact, to direct the wood to be damped is almost to convey an erroneous impression. At the most a little damp sawdust should be sprinkled on the hollow side, and allowed to remain for a few hours, for if the hollow side is swelled too much it will become rounded. The safest course, perhaps, is to lay the board with the hollow side down on a cold stone floor. Even though this is not perceptibly damp it will very often have the desired effect in the course of a few hours or a day or two. Sometimes it is even sufficient to lean the board against a wall, hollow side to it, or to lay two hollow-faced pieces on top of each other. Any of these courses may be tried, and if they do not answer there is nothing for it but to plane the wood level. Actually wetting the boards and then drying them under sufficient pressure to keep them flat is not a good plan.

As it has an immediate practical bearing on many details in construction, it will be well to bear in mind that wood does not contract in length but merely in breadth. This fact seems to be often forgotten, but as it will be found mentioned more in detail elsewhere nothing more need be said about it now.

It will be observed, as experience widens, that the value of timber may be looked at from two points of view—that of utility and that of cost. Thus pine, whitewood, and bay mahogany are valuable to the cabinet-maker from their good qualities, though not expensive; while others, though costly, either because they are fashionable or for other reasons, are from a manufacturing standpoint not worth so much because they are unreliable or difficult to manipulate. If, therefore, the cabinet-maker meets with any wood with which he may be unacquainted it does not follow that because it may be low priced that it is unsuitable, nor the converse if it is costly.

The question of waste is one which is of considerable interest to the cabinet-maker, waste being defined as those short pieces which are comparatively valueless. It is impossible to avoid this, so that on reckoning the quantity of timber to be got specially for any piece of work due allowance should be made for it. To buy simply the number of feet of stuff which the finished job contains would not be nearly enough in most cases. Thus, if a sideboard top measuring 6 ft. × 2 ft. = 12 ft. is to be made, to suppose that 12 ft. of wood only would be required would probably leave the matter very short. If the pieces are of exactly the length, viz., 6 ft., or of such lengths that they will cut into them, very little more, only enough to trim the edges, would have to be reckoned for, but this rarely happens, and the short ends must be to some extent considered as waste. Of course, if of any size they may be usable in other parts of the work, but in any case there must necessarily be a good deal of waste. In large factories it can be averaged, but the small user can hardly do this. The amount of actual waste must necessarily depend on circumstances which it is impossible to enumerate here, and it will be sufficient to direct the attention of those who make furniture as a matter of business to keep a careful watch over this item when estimating the cost of anything. To regard every short piece cut off as absolute waste—that is, worth nothing—would, of course, be unfair to the purchaser, but full value certainly should not be calculated for the 'short ends,' which will be caused by making up almost any article of furniture. In connexion with this part of the subject it may be well to suggest that as the pieces accumulate, for anything that is likely to be useful should not be burnt as is often the case, they should be put on one side, when many of them will answer for portions of other work. It is not an economical plan to cut into a fresh board for a small piece which may be required when something equally suitable can be selected from the 'short ends' corner.

CHAPTER IV.

GLUE AND ITS PREPARATION.

Frequent Use — Selecting Glue — Preparation — Employment — Preservation — Liquid Glues — Brush.

IN making up any piece of furniture, it is well known that glue is largely used, so largely, indeed, that the cabinet-maker cannot do without it, although there are theorists who object to it. Theory, however, is one thing and practice another, for the custom of using glue is so well established that no one interested in the manufacture of furniture could think of dispensing with it. As the use of glue is sometimes deprecated on the alleged ground of its instability, it may be said that glue, or, what is much the same thing, some kind of cement, has been used from time immemorial by woodworkers, except when it is obviously unsuitable, and in construction exposed to the weather. For ordinary furniture, glue is and may be used freely, though perhaps it may be well to say that it is not reliable in damp, tropical climates.

The cabinet-maker is strongly recommended to use only the best quality of glue, and not be tempted by low prices, for there is plenty of it made and sold which, even if it may be suitable for other purposes, is not fit for good and durable furniture. A very fair idea of the quality of glue may be got by noticing its appearance, and generally this will be a sufficient guide to the novice when purchasing. A very slight acquaintance will enable him to discriminate between good and bad, and the following hints will be of assistance in doing so. The principal feature, without going into more minute tests, which could only be appreciated by an expert, is the colour and transparency of the cake. It should be of a good clear brown or tawny tint when looked at by transmitted light. Inferior qualities are dark and muddy looking. Very light-coloured glues, as a rule, are not as strong as the clear brown, though they are occasionally useful for special purposes. For ordinary use, however, they are not so suitable, and the beginner may as well leave them alone. If he can, let him get the best English or Scotch glue, for there is nothing better, though some of French or German origin may look nicer. Foreign glues are not much in favour with English artisans, though the best qualities are infinitely superior to much of the black opaque stuff which is sold.

Not of less importance than the original quality of the glue is the way in which it is prepared and used, for a joint made with the best of it will be weak unless it is properly employed. This fact is often ignored by amateurs, and the contents of the ordinary domestic glue-pot are seldom satisfactory. Before proceeding further, it may be well to explain how to prepare the glue for use. As is well known, the recognised form of glue-pot consists of two vessels, the outer one containing water and the inner one the glue, which should never be heated in anything brought directly in contact with the fire. If it is, the glue is very apt to be burnt and rendered worthless. With water surrounding it, it is impossible to overheat the glue. Before, however, the glue can be melted by heat, it must be softened in water, and while doing this further notes of its quality can be taken. Sufficient should be broken up in small pieces and covered with cold water, in which it is allowed to stand for some hours till thoroughly soft. If the

GLUE AND ITS PREPARATION. 61

glue dissolves in the cold water it is poor. It should merely gelatinise and become soft without dissolving, and speaking roughly, the more water it will absorb the better its quality. Of course the water causes the glue to swell up, so that it should be well covered. When quite soft, the surplus water may be poured off and the glue be melted over the fire in the glue-pot, care being taken to keep plenty of water in the outer pot. When the glue is melted it is ready for use, and is at its greatest strength, for—and I particularly wish to impress this fact on the novice—the oftener glue is melted the weaker it becomes. This is the reason why household glue is so often defective and will not hold. The old glue is melted and remelted whenever it is wanted, till at last it has become worthless as an adhesive. It must not, however, be understood that glue should not be melted more than once, for if good originally it may be heated several times without apparent depreciation, though this is constant. For this reason, therefore, no more glue than can be used within a reasonable time should be prepared at once.

If the glue when hot seems too thick, as it very likely may be, it can easily be thinned by putting a little hot water to it. As that from the outer kettle is very handy, it should always be clean and not allowed to remain in impregnated with rust. The proper consistency of glue is of some importance, but it is impossible to give any definite directions about it. Perhaps the best rough test is to dip the brush into it and notice how it drops from it. If it does so in lumps or in a very thick stream, some more water is required; but if it runs like oil of medium consistency, it will be about right. It should be thin enough to be rubbed freely into the wood without being lumpy. A thick coating is not necessary, for all excess in a joint must be pressed out, as the object is to get the pieces of wood as close together as possible. There is no greater mistake than to suppose that the larger the quantity of glue in a joint the stronger it will be, for the reverse is the case.

Perhaps of equal importance to the freshness and consistency of glue is the necessity of using it as hot as possible and bringing the glued surfaces together while it is so, for the adhesion otherwise cannot be good. It will be noticed that as the glue cools, either in the pot or on the wood, it changes from a liquid to a stiff jelly. In this condition it may be kept for a considerable time if the air is excluded from it so that the moisture cannot evaporate. By taking advantage of this, the trouble of frequently waiting while the hard glue is soaking may be avoided by those who only use it occasionally or in small quantities. All that is necessary is to pour some glue while hot into a tin, from which sufficient for use can be cut out from time to time as required. This is the nearest approach which can with safety be made towards keeping glue in a usable condition, for the various methods which are sometimes mentioned for keeping it in a liquid state cannot be recommended; the best properties of glue are nearly always injured by such treatment. Leland, in his *Manual of Wood Carving*, says that if about a teaspoonful of nitric acid is added to half a pint of hot glue this will be improved as well as remain for a considerable time in a liquid state. I have not used this preparation, but name it, as the author who gives it is generally reliable, and the knowledge may be useful to some. Good glue prepared in the ordinary way, however, is satisfactory for all purposes of the cabinet-maker.

It should be mentioned that as the water evaporates from the glue in the pot it is necessary to add a little occasionally to supply the deficiency. In case some readers prefer to have a liquid glue which can be used at all times cold without any preparation, it may be said that Lepage's fish glues are very good, and have excellent adhesive qualities. They do not set or become hard so quickly as the ordinary kinds, so that they allow of a little more time being taken. As hot glue sets quickly, some more so than others, one that allows more time when gluing large surfaces is sometimes an advantage. I do not think even the fish glue is quite so strong as the ordinary kind when this is of the best, but it is at least equal to the average, and better than inferior qualities, and so far as I have seen is uniform. Unless bought in large quantities it is dearer than the hard kind, though as there is little waste with it the ultimate cost is perhaps not much greater.

In the absence of a proper glue-pot a very efficient substitute may be found in an empty jam-jar and small saucepan, anything indeed that will hold the glue and permit of its being melted in a water-bath.

It seems almost unnecessary to say that the glue is rubbed on the wood with a brush. Almost any kind will do, though by preference it should be moderately stiff, and of course in size proportioned to the work. A good useful brush may be made with a piece of cane. The hard, flinty skin is cut off for a short space at one end, and this is then hammered till the fibres are sufficiently loose.

CHAPTER V.

NAILS.

Screws—Sizes—Brass Screws—Brads—Wire or French Nails Needle Points—Dowels—Dowel Plates—Glass Paper—Stopping.

THE varieties of nails used in cabinet-making are not numerous, and there is no reason why those most commonly employed have the preference given to them, except that they are the most convenient forms. Any nails, therefore, which may suit the purpose may be used when they are more convenient than others.

The principal nails used are screws. These are often called wood-screws, though made of metal, in order to distinguish them from those used in metal working, which as regards thread and other important particulars are of different construction.

Wood-screws are to be had in several varieties, the sort principally used being of iron and having flat heads. They are also made with rounded heads and japanned, these latter, however, being seldom required. More useful than these are the brass screws, which are also made with flat and rounded heads. There is also another variety with the plain part of the shank in the middle and a screw at each end. These, which are known as double-ended or dowel screws, are not much used, but they come in handy occasionally for special purposes such as joining turned columns above and below a shelf. It is, however, quite possible to dispense with them and to use wooden pegs instead.

Screw-nails are made up in packets containing one

gross, and are distinguished by a number indicating the diameter of the shank, and by the length. Thus a No. 1 screw, whatever its length, will always be found of the same thickness. Lengths increase by $\frac{1}{8}$ inch from $\frac{1}{4}$ inch to inch, after that by $\frac{1}{4}$ inches, so that there is plenty of variety. Needless to say that it is quite unnecessary for the cabinet-maker to have all the sizes in stock for him to work from, as a few judiciously chosen will serve almost every purpose. So much depends on what the bulk of the work is that it is hardly possible to reliably advise every one what screws will suit him best, but the following will probably be found as useful an assortment as any for the general cabinet-maker. If he wants others he can easily procure them as occasion arises: — For general carcase work, Nos. 10 and 12, $1\frac{1}{4}$ inch and $1\frac{1}{2}$ inch. For odds and ends, locks, hinges, &c., No. 6 and 8, $\frac{5}{8}$ inch, $\frac{3}{4}$ inch, and 1 inch. The smallest and largest sizes made are seldom wanted.

The iron wood-screws do well enough for structural parts where they are not seen, and are often used exclusively. The appearance of any work is, however, better if brass screws are used whenever the nail heads are visible or when used to fasten on the brass-work, such as hinges, escutcheon plates, handles, &c., also for beadings holding door panels in the frames. The flatheaded nails are used whenever the head is sunk level with the surrounding wood or brass, as in hinges, the round head ones being better for beads, handles, and those parts in which it is not convenient to sink the heads. Only the smaller sizes of brass screws are required. For screwing on handles Nos. 2 and 6, $\frac{5}{8}$ inch are very useful. Those with flat heads cost from two to three times as much as the corresponding sizes in iron, while those with rounded heads rank equal in price with the next larger size in flat heads.

Brads are almost too well known to require any

F

remarks, except that they come in handy for many parts in which it is not necessary to use screws. They are slightly tapered, with a kind of hook or return for a head which allows of them being driven close in.

Wire or French nails have of late years largely superseded brads, and may generally be recommended in preference to them. They are to be had in a great variety of sizes, and when of brass and of small size they are often used to fix in the beadings behind panels, though they neither look so well nor are they so convenient as screws if from any cause the panels have to be taken out.

Tacks and large-headed nails are hardly ever required in cabinet-work.

Needle-points are sharp pieces of steel very like needles without eyes. They are often extremely useful for various purposes, though it is not easy to define them. It will suffice to say that from their small size and the facility with which they can be snapped off short at the wood they can be used in places where the presence of a nail of the ordinary kind would be objectionable. They are specially useful when veneering to hold the veneers down temporarily when nails could not be inserted. They can hardly be regarded as indispensable, and many, for some purposes, prefer thin wire pins, which seem to be taking their place. In many parts of the country heckle pins, which I understand are used in cotton manufacturing, are preferred for veneering purposes, but I do not think they are obtainable everywhere.

So closely allied to nails are dowels or dowel-pins that they may appropriately be classed together. In most places the user will have to make these for himself, as he very easily may. In large towns, where the *manufacture* of furniture is of any importance, they may be obtained ready, or perhaps I should rather

NAILS.

say partly so, for they generally want a little attention from the cabinet-maker before he can use them. Their use will be found described elsewhere, meanwhile their construction may more appropriately engage attention. It may be well to explain that they are merely pieces of round stick of various thicknesses, those most common being $\frac{3}{8}$ inch and $\frac{1}{2}$ inch. Short lengths are cut off as required, and as they are generally invisible after the work is made up, I may refer readers to about the only place in which they can be seen afterwards, viz., on the edges of the leaves of an extending dining-table. The short pegs which fit into holes in the corresponding part of the tables are dowels. In this case one end, that which is seen, is unglued, but for ordinary constructive parts both are secured with glue. From this it will be seen that a dowel may be described as a double-ended wooden nail. As they must fit very accurately, tightly, within the holes bored for their reception, they must be made of uniform size with the bit which is used for boring the holes, and this is how it is managed. A steel or iron plate, say $\frac{1}{8}$ inch thick, has a hole or holes bored through it, the edges being left sharp. This plate is either sunk into the bench top or left loose, or for reasons to be shortly shown, is mounted on a piece of wood an inch or so thick. Of course a hole must be bored through either this or the bench top at least as big as that in the iron. The dowel wood in any convenient length, say 6 to 12 inches, is roughly planed to shape, and is then hammered through the plate to finish it. Almost any hard straight-grained dry wood will do for dowels, the preference, if any, generally being given to beech. Now it will easily be seen that on driving a dowel into a hole which it fits exactly the air beneath it must be compressed, and to allow this to escape a channel must either be cut in the dowel, or a side of this be flattened with a rasp or something convenient. To take too much off is not advisable, and a channel can easily be cut while forming the dowel by having a sharp point projecting through the wood block above-mentioned into the hole. A screw-nail, with the point filed sharp, will at once suggest itself as being an easily formed cutter. It only remains to be said that dowel wood should be thoroughly, not merely nearly, dry before the dowels are formed. If at all damp they will be apt to shrink either before or after they have been used, and in either case will not fit so tightly as they should.

Glass paper, or, as it is often termed, sand paper, though, unlike the foregoing, it does not actually enter into the construction, is essential for preparing the wood of those parts which must be finished as smoothly and cleanly as possible. It is sold in sheets of 12 inches by 10 inches, and is prepared in various degrees of coarseness, any of which the user can select to suit his work. Nos. $1\frac{1}{2}$ and middle 2 are the most useful. It is used over a cork rubber of any convenient shape and size, and should on no account be simply pressed on the wood with the fingers direct. The blocks generally used are 4 or 5 inches long by 3 inches wide, and rather more than an inch thick, but there is no rule as to size except that it should be such that the cork can be conveniently grasped on top by the worker's hand, and the paper be held firmly wrapped over the edges. The edges of the lower surface should be slightly rounded off.

I must touch on a point out of deference to public opinion, erroneous though this may be. The supposed use of putty to fill up and disguise bad work is alluded to. I am far from saying that this material is never used, but it certainly would not be permitted in any respectable shop unless for filling nail holes in the commonest and roughest kind of furniture. A good cabinet-maker has no need to resort to such artifices to

NAILS.

hide bad work. Occasionally something is necessary to fill up natural flaws in wood, and then a stopping may be used. There is nothing better for the purpose than a mixture of ordinary resin and beeswax (wax alone should never be used) melted together. The proportions of each may vary considerably, but provided the mixture is hard enough nothing more is requisite. It may be rolled out before it is cold into sticks, in which form it is most convenient, but the more common way is just to melt a little, as required, in an iron spoon or anything handy, and work it into the fault with a piece of hot iron, scraping and papering it down smooth when cold and hard. Besides tools and various appliances there is little else that is used by the cabinet-maker in the shape of material.

CHAPTER VI.

TOOLS.

Selection and Care — List of Tools — Saw Teeth — Panel Saw — Sharpening and Setting — Tenon Saw — Dovetail Saw — Bow Saw and Frame Planes — Iron and Wooden Planes — Plane Irons — Jack Plane — Trying Plane — Smoothing Plane — Rabbet Plane — Plough — Old Woman's Tooth — Hollows and Rounds — Chisels — Gouges — Spokeshave — Gimlets — Bradawls — Brace and Bits — Gauge for Bits — Screwdrivers — Marking, Cutting, and Mortise Gauges — Pincers and Pliers — Square — Bevel — Hammer — Mallet — Punch — Compasses — Rule — Scraper — Scraper Sharpener — Marking Awl — Grindstone — Cork — Handscrews — Holdfast — Cramps — Files — Dowel Plates.

THE subject of tools is one of the utmost importance in any handicraft, and the workman cannot exercise too much care in their selection, both with regard to quality and variety.

There is much truth in the old sayings that 'A bad workman always complains of his tools,' and 'A good worker can use anything.' The former individual, however, seldom has good tools, or if they were so originally have been allowed to fall into bad condition. This is as much as to say that a bad workman is careless with his tools. It may be quite correct that a good workman can do with anything or with poor tools, but it is equally correct that he seldom does use such, for he takes a pride in having them of as good quality as they can be got, and in keeping them in good order. It is hardly too much to say that the quality of a man's work can be estimated by his tools — not by their number or magnificence of appearance, but by their condition. They may be worn and dirty-looking from use, so far as the handles and other secondary parts are concerned, but their edges are sharp and orderly. The amateur, I know, is very much taken with fancy tools, polished handles, and all that kind of thing, and such an appreciation of the beautiful may be all very well, but it is of far more importance to pay attention to their working qualities. If they are not adapted to their purpose they are of little value, however nice they may look. It is no use paying for mere appearance.

The novice must also be cautioned that the number and variety of tools in his possession do not necessarily make a man a good workman. A few tools, properly understood and used with facility, are of infinitely greater value than a large number which cannot be used with precision. Those who have plenty of money and do not mind the outlay, may, of course, lay in enough tools to stock a dealer, getting one of everything to begin with, but a much wiser and better plan will be, after purchasing the few tools which may be looked on as indispensable for a start, to add others as they are required. I suppose there never was an amateur who went in for, as he thought, a complete outfit at the beginning but found, in the course of time, that he wanted something more, and that many of the things he had got were seldom or never used. Men's tastes in regard to tools vary as much as in other matters, and the tool which one considers indispensable for any particular detail of work another may think of no great value. Then it must be remembered that special tools for special purposes often require a considerable amount of manipulative ability before they can be used with advantage, or with an appreciation of their value. The tool which in experienced hands might effect a saving of time and do the work in the neatest manner, might, to the beginner, be useless. It is, therefore, **not** necessary for him to get many things to start with, and the outlay need not be a heavy one. If I may venture on somewhat dangerous ground, I would say to the beginner that it is part of the dealer's business to recommend as many tools as possible, so that the intending purchaser should at least have some idea of what he really wants.

As has been said, the tools should be of good quality to begin with, and this does not necessarily imply the payment of the highest prices, for really useful, though not so highly finished, things can often be got at very moderate figures. In all cases it is best to get tools from a regular tool-dealer, and not from fancy shops where everything is dealt in. The cheap sets, made up and sold in the latter for amateurs, should be avoided, as they are little better than toys, and satisfactory work cannot be done with them.

The novice may be inclined to ask or expect some directions how to distinguish between good tools and bad. The best advice I can give him is to buy only from respectable dealers, of whom there are plenty, and to trust to their judgment about quality rather than to his own. The prices for equal qualities of tools will not be found to vary much wherever they may be bought, though occasionally, in small places where there is no competition, they may be higher than in large towns. In such cases the purchaser can always protect himself by buying from good London dealers, most of whom make equitable arrangements for carriage. The well-known house of Moseley & Son, for instance, pays carriage to any part of England on parcels over ten shillings in value, while Lunt allows a discount instead.

Care should be taken of tools when they are got. They should be kept in good working order, and not allowed to get rusty, as they soon will if left to lie about in a damp place. The amateur, using them only occasionally, should be particularly careful in this respect, though the remark is equally applicable to the professional cabinet-maker. He, however, may be supposed to have his tools in such constant use that any approach to rust on them will be checked at once. If tools must be allowed to remain unused in a damp place, the amateur should wipe them over now and then with a little oil or grease of some kind.

The edges of tools which are in use require constant attention, and should be sharpened as soon as they get at all dull; good work cannot be done with blunt tools. It is a bad habit with some amateurs, and one not quite unknown in the practical workshop, to allow all tools of a kind to get among the 'middlings' and then sharpen them up altogether. It is far better to attend to each as soon as it loses its keen edge, and then the trouble of sharpening is hardly noticed, for an occasional rub on the oilstone does all that is necessary.

The following list contains those tools which may be considered indispensable to the cabinet-maker, though it must be remembered it is by no means an exhaustive one, for there are many other things which it may occasionally be an advantage to have. It is, however, well to point out that all of the special tools which are sometimes used are merely modifications of the commoner varieties; a little ingenuity, therefore, will often enable these to be used instead of something else. In case it should be thought that the list is a meagre one, I may as well say that the object is to show the worker of limited means how he may economise and to make the art of cabinet-making as simple as it can be for the novice, to help him instead of to mystify and discourage him:—

	s.	d.		s.	d.
Panel Saw, 24 ins.	4	9	Jack Plane, 17 ins.	4	6
Tenon Saw, 14 ins.	4	0	Trying Plane, 22 ins.	6	0
Dovetail Saw, 8 ins.	3	9	Smoothing Plane	4	0
Bow Saw, 12 ins.	3	6	Rabbet Plane	2	6

Item	s.	d.	Item	s.	d.
Plough Plane and 8 Irons	17	0	Scraper	0	6
Hollows and Rounds, 9 pairs	15	0	Marking awl	0	2
1 doz. Chisels, firmer or paring	7	0	Oilstone	3	6
½ doz. Gouges	4	0	Scraper Sharpener	0	3
3 Gimlets	0	9	Cork Rubber	0	6
3 Bradawls	0	6	Mortise Chisel	2	
Brace and Set Bits (36)	16	0	Toothing Plane, 2 Irons	3	
2 Screwdrivers, large & small	2	6	Old Woman's Tooth	1	
Mortise Gauge	2	0	Handscrews, 9-inch (4)	3	0
Cutting ,,	0	9	Holdfast	5	6
Marking ,,	0	6	1 Iron Cramp	10	0
2 Spokeshaves	1s. & 1	6	*2 Wood Cramps		
Pincers	1	0	Mallet	1	0
Square	1	6	Dowell Plate, 5 holes	2	6
Bevel	2	6	,, Bits to match, at 1s. 3d.		
Hammer	1	6	Cutting Pliers	1	6
Punch	0	2	Wood Rasp, 8 in. ½ round	1	0
Compasses	1	3	File, 8 in. ½ round	1	0
Two-foot Rule	1	0	*Wood Square, useful for panels	–	

* To be made by user.

The prices are named to give some idea of cost, and are not intended to be taken as definite, for in many instances tools are to be had at lower figures; nor must the purchaser, on the other hand, think he is necessarily being imposed on because he may occasionally pay more.

With such a stock of tools the novice at cabinet-making may consider himself fairly set up; and at first he need not even buy all those that are mentioned if he does not require them. Thus, he might do with fewer chisels, and less than a complete set of bits. In addition to these tools there are several other appliances which are necessary, such as cramps, &c., all of which will be found described in due course.

To give the novice some idea of the tools in the foregoing list they may be mentioned more in detail, with such hints on their use as may seem desirable at this stage.

Saws.—On looking through any tool catalogue the novice may be bewildered by the number he sees named. If he asks advice about the different sorts he may be told they **are** all useful if not absolutely necessary. No doubt they are so in certain circumstances. It all depends on the kind of work to be done. Without at all pretending to write a treatise on saws and their varieties, it will be well for the novice to understand something about the principles on which they are made. The serrated or toothed edge is, of course, well known, but the minuter differences which exist have probably not been regarded, though any one who has ever used saws, even occasionally, must have noticed that some cut more easily than others. It does not, however, follow that because a saw cuts well in one material it will be equally

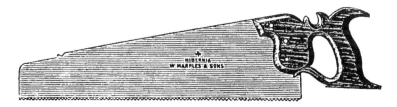

Fig. 1.—Handsaw.

good in any. Thus there are saws which are specially adapted for cutting thick, soft wood with the grain or 'ripping,' while others have a different form of tooth, which makes them more useful for cutting hard, thin wood across the grain. The cabinet-maker has little use for the former kind, as most of his work is in comparatively hard wood, while there is so little ripping to do that he can very well dispense with a saw of the best kind for that purpose. All the same, he wants one with which he can manage any ripping which he may have to do, and if he is going in for several he may as well get a rip saw among them. For ordinary work, however, he can manage very well with a 'half-rip' or 'panel' saw, the teeth of which may not be so good as some others for cutting pine with the grain, but will do very well. As a good deal depends on the size as well as the shape of the teeth, let it be said that a handsaw, the general form of which is shown in Fig. 1, with about six teeth or points to the inch, is what is wanted. The teeth should be pointed as in Fig. 2, and not as in Fig. 3, though these are best for ripping solely.

Fig. 2.—Size and Shape of Saw Teeth.

Fig. 3.—Rip-saw Teeth.

A good handsaw should be flexible, so that when bent it will resume its flatness, and any bend observable would at once proclaim it to be defective. On looking at the teeth it will be seen that they are slightly bent alternately to each side. This bending is termed the setting, and is in order that the friction between the plain surface of the blade and the wood may be minimised, so that the passage of the saw is made as easy as possible. Too wide a set would, however, cause too wide a cut or 'kerf,' so that, especially when cutting thin stuff across the grain, the saw would wobble and cause an unnecessary amount of strain in keeping it straight and to the line.

It will also be noticed that the teeth, instead of being filed straight across the blade, at right angles to it, are slightly on the slant. This is to give each tooth as much as possible of a cutting edge, for if filed straight across the action would be entirely a tearing one, and the labour of sawing would be thereby greatly increased. Naturally in course of time the points get blunt, and the saw requires resharpening. This the novice may do for himself, but if he is wise, and there is a tool-shop or some one handy who is accustomed to this kind of work, he had better get it done for him. The sharpening and setting of saws is an art which is not picked up in a day. Sometimes, however, the user may be so placed that it is a case of Hobson's choice whether he does the work himself or puts up with a blunt saw, and the following hints may be useful to him. A triangular saw file will be required. Instead of occupying space with figures and diagrams showing at what angle to work the file, it will be better to direct the worker to notice how the teeth have been shaped, and to follow in the old lines as closely as possible.

First of all, a file should be run lengthwise from end to end of the saw to reduce all the teeth to the same level. They are then to be filed to get the points sharp. In doing this, care should be taken to have them equally filed, for of course if any of the cutting points project beyond the others, they will have to do most of the work, as those set back will to some extent be inoperative. The setting of the teeth sideways is usually done by hammering alternate teeth over the edge of a block of iron, then reversing the saw and treating the other side in the same way. Perhaps in expert hands this is the best method of setting, though opinions differ on the point. Without discussing the *pros* and *cons*, it may, however, be admitted that the novice at saw-sharpening will find the use of one or other of the saw-sets which have been devised of considerable assistance to him. A great deal depends on the setting, so that it must not be done in an irregular manner, or it will be useless to expect the saw to work properly. To save repetition, it may be said that all saws are sharpened in much the same manner, the chief difference between the different kinds being in the shape, size, and set of the teeth. It may also be well to note that there is no abrupt transition from one kind to another, the differences in some instances being so minute that saws of different names almost merge into each other.

This is the case with the other two named on the list, the tenon and the dovetail saw, for they are practically the same things, the principal difference between them being that the latter is smaller and finer than the other. Both of them are used for a variety of purposes other than those indicated by their names, and the maker of small things will hardly need anything larger.

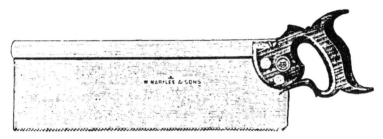

Fig. 4.—Tenon Saw.

The tenon saw is shown in Fig. 4, and the dovetail is similar in appearance, except that its handle is usually open at the bottom. Both these saws are made of thin

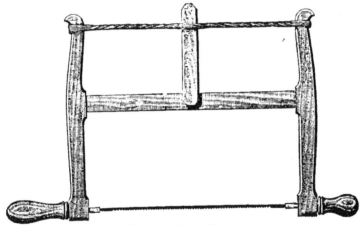

Fig. 5.—Bow Saw.

metal, and are strengthened by a rim of iron or brass on the back edge; those with the latter being the more costly, but not necessarily better for work.

The bow saw—or, as it is often called, the turning saw—is shown by Fig. 5. Properly speaking, this is a composite tool, as the saw and frame are separable portions, though neither can be used without the other. The blade is between the two handles, and is kept at the requisite degree of tension by the cord and lever at the back. As the handles are movable, the saw can be turned with its cutting edge in any direction. The use of this tool is obviously to saw curves, which could not be done with a straight, wide saw of the ordinary kind. The blades can be obtained separately from the frame when needed, and are made in various lengths, to any of which the frame can be adapted by merely altering the bar between the string and the blade.

Planes.—These tools are used for a variety of purposes, smoothing wood, both on the surfaces and edges, cutting grooves, forming mouldings, and suchlike work, according to the shape of the cutting iron. This may be compared to a chisel, but instead of being loose, so that it can be used in any position or at any angle to the wood being operated on, it is held firmly in a block. This serves as a guide to keep the edge from entering the wood wrongly. The chief use of the plane is to smooth and level, or, perhaps, it should rather be said that more wood has to be made level than have any other form given to it. The primary planes have, therefore, a straight cutting edge and a corresponding flat sole, while the others have the edges and soles shaped to suit any work they are intended for.

The ordinary English plane has a wooden block usually of beech, but of late years many iron planes have been introduced, chiefly from America. It is sometimes considered that these latter are superior to the former in every way, and as the question is sure to occur to the novice whether he should use the iron or the wooden plane, a few remarks are not uncalled for.

I am not prepared to advocate the decided advantages of either kind over the other so ardently as some do, the advocacy of those who generally reach the public ear being generally more in favour of the newer-fashioned iron planes. In the practical, everyday life of the workshop, however, the old wooden plane holds its own; and whatever may be the case in other crafts, the cabinet-maker as a rule works with it. It is not at all an uncommon thing to find that a man who has iron planes, which he has got either from curiosity or with an idea that they may suit him better than the others, after a time discards them altogether or uses them very occasionally. Here and there a man may be found who uses iron planes exclusively, but he is the exception, and in any cabinet workshop the wooden plane holds its own. Of course, in saying this I must not be understood to imply that iron planes may not be used by those who prefer them, but I think it is only fair to the novice that he should not be misled into supposing that better work can be done with iron planes than with wooden ones, or that the former are necessary to him.

Equally good work is done with both. In theory perhaps the iron plane is better; in practice the wooden one, when properly used and understood, is just as good and often better. By many it is assumed that the iron plane must work more truly, be more durable and less likely to get out of order. These are specious arguments, and if nothing were said about them the novice might conclude them to be unanswerable. The quality of the work done with wooden planes is quite as good as that with the others, except in very rare circumstances. Iron is of course more durable than wood, but then there is not a vast amount of wear on the sole of an ordinary plane. A good wooden one will last a man a lifetime if proper care be exercised, so surely nothing more can be necessary. An iron plane may be faulty or defective when new, just as easily as a wooden one, and a defect in it is not so easily remedied. Wooden planes are to be had quite as accurately made as any in iron, and if properly treated they will remain in perfect condition. In cost there is no comparison between the two, the advantage being decidedly in favour of wood, to which the prices named on the list exclusively refer. The low-priced, toy-like planes of iron hardly come within the scope of comparison, as they are not used by the practical artisan. A wooden plane which has been very extensively used may get so worn that the mouth becomes too wide for some purposes. In such a case it is a comparatively easy matter to let a new piece into the sole or to resole entirely either with wood or iron. I cannot, however, too strongly advise the novice not to tamper with the mouth or sole of any plane he may have. Directions for rectifying are unnecessary, for the simple reason that by the time any truing up is required the user will be a novice no longer, and may safely be left to his own devices for doing the needful rectification. Let him get good planes to begin with, and if after a time he fancies they are unsatisfactory he should get some experienced hand to pass an opinion before deciding that they are defective. They may be so, for even the best of tools will show unsuspected faults occasionally, but the novice certainly should not attempt to rectify them. If he does, he will probably increase the fault or cause others. My advice to the novice and amateur is to get good wooden planes, and then when he can use them to full advantage to get the more expensive kind if he thinks they will be better. It should be said that iron planes are occasionally of English make. In addition to those planes which have blocks either all iron or all wood, some varieties which have principally the latter are fitted with iron soles, when for all practical purposes they may be treated as

G

iron planes. If these latter have any advantage over the wooden ones, it may be said to be principally when planing end grain or a particularly cross-grained piece of stuff. In competent hands, however, a wooden plane will do anything that a plane can do.

On purchasing a new plane it is advisable to thoroughly oil the block. This may best be done by immersing it in raw linseed oil and letting it lie in it for a week or so till soaked. Such treatment will cause the plane to work more sweetly than it otherwise would, and by filling the pores with oil will lessen any danger of the block casting or becoming untrue on the sole. On examination of the ordinary bench planes—jack, trying, and smoothing—it will be seen that the irons are double. The larger of the two is the true cutter; the other, the break iron, or back iron as it is often called, though it actually is in front of or on the other, being used to turn up the shaving and prevent the wood being torn. On the width of the mouth, as the opening across the bottom of the plane is called, and the distance of the edge of the back iron from the cutting edge, the thickness of the shaving depends. The width of mouth, of course, remains fixed, except from the widening by wear, while the cutting iron can be regulated. For coarse, rough work, as that done by the jack plane, the mouth may be comparatively wide, and the edge of the back iron be set one-eighth of an inch from that of the other; while for fine work, as with the smoothing plane, the mouth is narrow and the cutting edge only very slightly in advance of the other. In any double iron plane, the nearer the two edges are together the finer will be the shaving, but the labour of planing will be increased. From this it will be seen that the relative positions of the two irons is of considerable importance, and that within certain limits the planes can be regulated to suit the work on hand. As jack, trying, and smoothing planes are made with both double and single irons, it should be said that the latter, though cheaper, are not suitable for cabinet work. With this we may proceed to consider planes more in detail.

Jack Plane.—This is a tool shaped as shown in Fig. 6, and is the one used first on a rough piece of wood, or when it may be necessary to reduce the size considerably. Either this planing down or the removal of roughness might be done with another and finer plane, but then there would be more labour and risk of injuring the finer tool and unfitting it for its special work. The

Fig. 6.—Jack Plane.

jack plane, then, is the one with which rough planing is done, and is used both to save time and the other planes which come after it. The usual length is about seventeen inches, and the width of the iron two and a quarter inches. The back iron, as only coarse, rough shavings are required to be taken off, should be set back about one-eighth of an inch. I may here venture on a piece of personal experience, from which those who can read between the lines may derive a useful lesson. An amateur complained that his jack would not work easily, and that nothing more could be done with it than with the trying plane. The explanation was not far to seek, for it was found that the edges of the back and cutting irons were close to each other. He knew this was right for a smoothing plane, and concluded it must be so for the other, forgetting that the two tools are for different kinds of work. To be scientifically correct, the angles at which the irons should be ground and placed in the stocks would vary, but in practical work this is very little regarded; in fact, I may say not at all in the workshop. The ordinary common pitch of the iron does very well for all the purposes of the cabinet-maker, and as for the angle at which the tool is ground, no special care is taken to get it scientifically correct. A good wide bevel is what is best, and a suitable one is shown in Fig. 7. The difference between it and that shown in Fig. 8 will easily be recognised. The bevel, it must be observed, is on the hinder or lower surface of the iron, the back iron having its edge in the opposite direction, so that when placed together they are as in Fig. 9. The grinding does not put a cutting edge on tools, for this must be done with

Fig. 7.—Angle for Plane Iron Edge.

Fig. 8.—Unsuitable Edge.

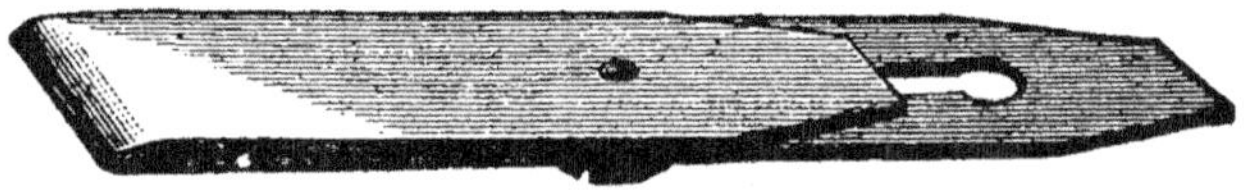

Fig. 9.—Plane Iron.

the oil-stone, and is only referred to here to say that after a plane iron or any cutting tool has been repeatedly sharpened, the edge gets as shown in Fig. 10, and requires regrinding occasionally.

Fig. 10.—Edge requiring regrinding.

Now a plane iron must be carefully ground, so that the edge is on a line with the sole. If it is on the slant, one side of the iron will project more than the other, with a corresponding difference in the thickness of the shaving removed. The iron will dig into the wood at one corner, and perhaps not act at all at the other. The work in such a case is bound to be defective. Though the sole of a jack plane may be flat, it is better to give the iron a very slight curve than to have it straight. Absolute smoothness is not got with this plane, and the slight convexity given to the blade facilitates the work. The curve, however, should be of the slightest, so that the whole width of the iron may act. If it is too great the iron will project so much towards the centre that it might be impossible to plane, unless the corners are within the mouth and consequently inoperative, in which case a cutting iron of, say, two inches, would not accomplish more than one of considerably less. To prevent the novice rounding the iron too much, perhaps it will be as well to suggest that the aim should rather be to get the corners slightly rounded off, as in Fig. 11, and only a very slight curve from corner to corner. Sharp corners and a straight edge do not answer in a jack plane. The irons are kept in position in the block by a wedge. This should be knocked in sufficiently firmly to hold the irons, which are kept together by a screw, firmly in their place. To loosen the wedge and remove the irons the plane is struck on top in front of the opening.

Fig. 11.—Curve of Jack Plane Iron.

The *Trying Plane* in general shape is similar to the jack; indeed, one may be used for the other, if kept specially for the work, and is shown in Fig. 12, where it will be seen that the principal apparent difference is in the shape of the handle. The length of the stock is, however, greater than the other, 22 inches being generally recognised as suitable. The edge of the iron should be straight across and the corners square, or only just so slightly rounded as to prevent them catching in the wood. This plane is used to level the inequalities left by the jack and to get the surface true. The back iron should be set closer to the edge, as large shavings are not wanted. Very closely akin to the trying plane is the jointer, which only differs from it in length. This is usually from 24 to 30 inches. Though it may be a useful tool occasionally for long, straight edges, it is by no means indispensable to the cabinet-maker, many of whom never use one.

The *Smoothing Plane* is used whenever a very fine, smooth surface is wanted. It is a much smaller tool than the others, and is of a somewhat different shape, as shown in Fig. 13. As the smoothing plane is only used for fine work, it should be kept exclusively for this and not be used on coarse rough wood. The blade should be ground straight across like that of the trying plane, and the edges of the two irons be very close together. To remove the wedge this plane

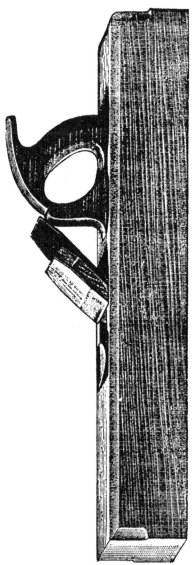

Fig. 12.—Trying Plane.

is knocked on the back a little above midway between top and bottom. A smoothing plane with a wide mouth is comparatively useless. Either an entire iron sole or merely a front can be added to a wooden plane at any time, being supplied by many dealers. The novice is not advised to fasten them on himself.

Rabbet or *Rebate Planes*, as shown in Fig. 14, are used for cutting away the edges of wood and forming a rectangular hollow. This work might be done with a chisel, but the difficulty of getting the depth and width regular would be much greater than with the special plane, which is neither more nor less than a block of wood holding a specially adapted chisel-like blade in one position. The iron, which is single, is at the lower end of the whole width of the sole. The block is of equal thickness throughout, and it will be noticed that the shavings as they are formed escape by an opening cut through from side to side instead of on top. The width of the iron varies, but one inch is a good, useful size. The mouth is either square across or skew, and it does not much matter which, though as the latter is better if anything than the other on hard wood and cuts soft equally well, it is, perhaps, preferable. The corners of the iron should not be rounded.

Fig. 13.—Smoothing Plane.

Fig. 14.—Rabbet Plane.

The *Ploughing Plane* (Fig. 15), or, as it is generally called, merely the Plough, is used for cutting grooves. As these are required of different widths, a set of eight irons is sold with each block, and are adaptable to it. By means of a screw on top the depth to which the irons will plough is regulated, as it is often of importance to get a number of grooves cut alike. To ensure the groove being straight, a movable fence forms part of the tool.

Fig. 15.—Plough Plane.

Toothing Planes in general appearance are similar to smoothing planes, but the iron instead of being at an angle is set vertically. The action of this tool is entirely a scraping one, and its principal use is to roughen surfaces when necessary to cause the glue to hold well to them, and to scrape down veneers on which a smoothing plane could not be used. The edge consequently is toothed very much like that of a saw, only instead of being filed they are formed by the front of the iron having a number of triangular grooves up and down its surface, so that on the blade being ground or sharpened these form a series of angular teeth on the edge. It is well to have two irons, one with coarse and the other with fine teeth, for different kinds of work.

The *Old Woman's Tooth* may be bought ready made—I hope no amateur will think I am taking 'a rise' out of him in suggesting this as a cabinet-maker's tool, for it is one in spite of its peculiar name—but there is no reason why the user should not make it for himself.

Essentially it is, or may be described, as a plough, but one of the crudest form, and little more than a block of wood with a hole through it in which a chisel or other cutting edge can be fastened by a wedge. It is used for cutting 'grooves or rabbets which could not well be managed with a plough or rabbet plane ; as, for instance, when the groove or rabbet is stopped short and does not run from end to end of the wood. Its action is somewhat rough, but it speedily removes waste wood to a uniform depth, this being regulated by knocking the iron through the stock. There is no universally adopted size or pattern, as being so often a 'homemade' tool the user forms it according to his own fancy. Fig. 16 shows as useful a shape as any. The edge of the iron should be as near the front of the block as convenient. The hole shown may be omitted.

Fig. 16.—Old Woman's Tooth.

Hollows and *Rounds* are planes something after the same style as a rabbet plane, but instead of having a flat sole they are either rounded or hollowed, and, of course, their irons are shaped to correspond. Their use is for working mouldings, and if only very plain furniture is to be made, or if, as is often the case now, mouldings are worked by power instead of by hand tools, they may almost be dispensed with. The young cabinet-maker working in a shop where steam or gas power is used will not have much need of them, though it is always an advantage to be able to form one's own mouldings. These planes are sold in pairs, of which sixteen form a full set, but a half set or eight pairs, made up of alternate sizes, will be sufficient for all ordinary requirements. In grinding and sharpening the irons great care must be taken to preserve the sweep, and not to alter it from that of the sole. To sharpen the hollows rounded slips of stone are required.

Chisels.—These, whether called firmer, paring, or mortise, are much the same thing, and between the former two the cabinet-maker need hardly distinguish, as the chief difference between them is that the paring-chisel is longer than the ordinary firmer. Perhaps a better way of putting it to the novice is to say that a long firmer is a paring-chisel, or that a short paring-chisel is a firmer. The latter often have thin or bevelled edges, as in Fig. 17, instead of as shown in Fig. 18, which represents the ordinary firmers, though these are also made with bevelled edges. The number named on the list may seem a large one, and those who do not care to get them all at once may buy them separately, as required. They vary in width from $\frac{1}{4}$ in. to 2 ins. The most useful sizes to begin with will probably be $\frac{1}{4}$, $\frac{3}{8}$, $\frac{1}{2}$, $\frac{5}{8}$, $\frac{3}{4}$, and 1 inch, others being got as wanted. The mortise-chisel has a much thicker blade than the others, as will be seen from the side view of one in Fig. 19, and the handle is better for knocking. The most useful sizes will be $\frac{1}{4}$, $\frac{3}{8}$, and $\frac{1}{2}$ in. It may be well to remind the amateur that other chisels should not be knocked with hammer or mallet; and if his workroom is accessible to other members of his household, it will be better to keep his chisels locked up. The ordinary domestic, and women generally, often seem to have a difficulty in distinguishing between a screwdriver and a chisel, but with strange perversity frequently select the latter to do the work of the former in sundry household jobs, such as raising carpets, opening boxes, and so on. The chisel is not improved by such treatment, and generally wants attending to with the grindstone afterwards. N.B.—It generally seems the best chisel which is used for such purposes.

The *Gouge* may be described as being a chisel with the blade rounded, as in Fig. 20. Though not so much used as chisels, they cannot be dispensed with, and a few should form part of the outfit. Unless of a very flat kind, when the edge of the oilstone may be used, slips of sharpening stone are necessary to put a good edge on them, and they are best with any kind,

Fig. 17.—Paring Chisel.

Fig. 18.—Firmer Chisel.

Fig. 19.—Mortise Chisel.

TOOLS.

Spokeshave.—The ordinary form of this is shown in Fig. 21. It consists of a small, narrow blade fitted into a long handle, and is used for planing or shaping purposes on curved surfaces where a plane could not be worked. In the ordinary spokeshave the blade is held in its place by two pegs, part of the blade, which project

Fig. 20.—Gouge.

through the upper part of the stock, and are consequently not seen in the illustration. If two are got, one should be large and the other small, but if only one,

Fig. 21.—Spokeshave.

the latter will be the more useful of the two. There are many 'improved' spokeshaves, but the old form holds its own.

Gimlets are well known as convenient boring tools, and it is unnecessary to say anything about them except that there are two principal varieties, known as the shell and the twist gimlets, which are represented by Figs. 22 and 23. Either kind may be used, some preferring one and some the other. They are not required in large sizes.

Bradawls.—These simple boring tools are also well known. The holes made by them result from compression of the wood, hence they are not convenient for hard wood. The blade beyond the handle is merely a straight piece of steel sharpened at the end on both sides somewhat like a screwdriver, for which the larger sizes may often be substituted when using small screws. As the end acts as a wedge, in boring holes with a bradawl, care must be taken not to run the risk of splitting the wood. This may be avoided by keeping the edge across the grain. With the ordinary bradawl, if only the handle is held when attempting to withdraw it the blade sometimes remains in the wood. This may be almost entirely avoided by keeping the forefinger on the blade. There are also some bradawls with improved ferrules which prevent the blades drawing out, but they are not obtainable

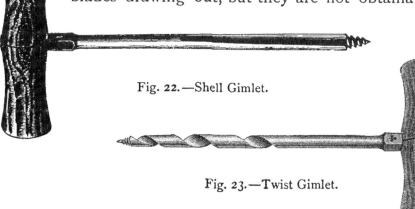

Fig. 22.—Shell Gimlet.

Fig. 23.—Twist Gimlet.

everywhere. Such a bradawl is the subject of the illustration, Fig. 24.

Fig. 24.—Bradawl.

The *Brace and Bits* are used together for boring holes which either from their size or other circumstances could not well be managed with a gimlet. The brace itself is nothing but the handle into which a bit may be fitted, and is made in quite an extensive variety of patterns, varying considerably in price. One of the most convenient forms is that illustrated, Fig. 25; but there are many who prefer the heavier-looking and more old-fashioned wooden pattern. In principle they are nearly all alike, the principal points to be observed in selecting one being ease of fastening and releasing the bit and

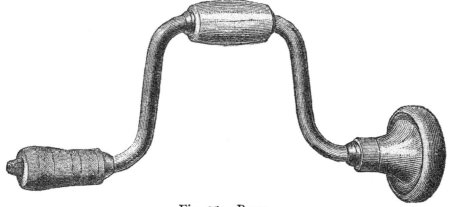

Fig. 25.—Brace.

easy turning. A brace which works stiffly is a nuisance. Though often sold with the bits, the brace may be got separately.

The bits may also be obtained separately. The complete set is considered to consist of 36, all of which will probably come in useful for some purpose, but smaller sets of 6, 12, 18, or 24 may be had. The entire set (36) may be had for 11s. black; if bright or straw-coloured they are more expensive without being any better, except in appearance. Many men consider the black bits superior to the others.

Probably the centre-bit, as shown on Fig. 26, is more used, or comes in for a greater variety of purposes than any other, as holes of a considerable size may be bored with it. The middle point acts as a guide, the other scribes the circle, and the bevelled edge opposite to it removes the wood. These bits, like others, are made in various sizes, but by using an expansion centre-bit one may be used to bore any sized hole. The best known is Clark's patent, one size of which bores anything between ½ in. and 1½ ins., and the other from ⅞ in. to 3 ins. They are expensive, about 7s. and 10s. respectively. A

Fig. 26.—Centre Bit.

much cheaper and equally serviceable, but not so well known, centre-bit is Anderson's patent.

To enumerate all the bits supplied is quite unnecessary, as most of them are for boring, and are shaped in different ways to suit different kinds of work. The best way for the novice to become familiarised with their special qualities will be for him to practice on pieces of waste wood and notice the results. Some, it will be seen, cut more cleanly in some circumstances than in others.

Some of the bits, it will be noticed, are specially adapted for widening the mouths of holes, or bevelling them to fit them for screw heads. Those intended for wood should not be mistaken for those specially adapted for metal work. It will be found necessary occasionally to widen the screw holes in various brass fittings. The screwdriver-bit is often very useful, as greater power can be exerted by means of the brace than with the ordinary screwdriver.

A very useful little appliance, whenever it is necessary to bore several holes to exactly the same depth, is shown in Fig. 27. The gauge is there represented attached to a bit than which none is better for boring dowel holes. It can be fitted to any sized bit, and regulated to a nicety. The mode of its application is so clear from the illustration that nothing need be said about it.

Screwdrivers.—Little need be said about these tools, as their use and general shape are well known. There are several patented forms, but apparently they are principally used by amateurs, as they are rarely seen in practical shops. In quite inexperienced hands they may have some occasional advantages over the common kind, one of which is shown in Fig. 28, but with a very small amount of practice these are quite satisfactory. In one of the best and most practical patent screwdrivers the screw head is held to the edge of the blade, and so might seem to supply a want. It is, however, seldom seen in the hands of practical artisans, and the ordinary form holds its own.

Gauges.—These are known as the cutting, marking, and mortise. In principle they are alike, and consist of a sliding block which can be fastened at any part of a bar, near one end of which a small steel blade or point is fixed. The object of the marking gauge, the scriber of which is little more than a piece of sharpened wire, is to scribe straight lines parallel with the edge or end of a piece of wood. In the cutting gauge, the cutter, instead of being a mere point, is a thin flat blade, set across the stock, and adjustable to various depths. It marks a finer line than the marking gauge, and may, therefore, often be substituted for it with advantage. It is a very useful tool for cutting through thin wood instead of sawing it. A common form of both marking and cutting gauges is seen in Fig. 29.

The mortise gauge has two cutters, which are adjustable at various distances from each other and from the block. It is useful when two parallel lines have to be scribed, as when marking for the joint from which it

Fig. 7.—Twist Bit and Gauge.

Fig. 28.—Ordinary Screwdriver.

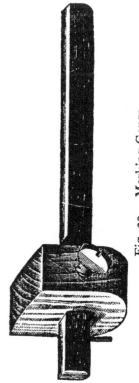

Fig. 30.—Mortise Gauge.

Fig. 29.—Marking Gauge.

H

takes its name; but it is possible to do without it and

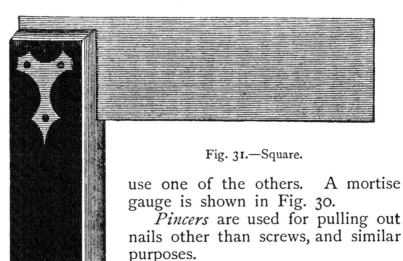

Fig. 31.—Square.

use one of the others. A mortise gauge is shown in Fig. 30.

Pincers are used for pulling out nails other than screws, and similar purposes.

Pliers, though not indispensable, will often be found useful, either for cutting wire nails, &c., or as small pincers.

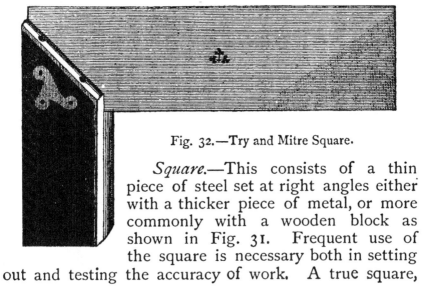

Fig. 32.—Try and Mitre Square.

Square.—This consists of a thin piece of steel set at right angles either with a thicker piece of metal, or more commonly with a wooden block as shown in Fig. 31. Frequent use of the square is necessary both in setting out and testing the accuracy of work. A true square, therefore, is of the utmost importance. A good useful size is the six-inch. By the use of a combined try and mitre square, as shown in Fig. 32, either squares or angles of 45° can be regulated.

Sliding bevel.—This is also a setting out and testing

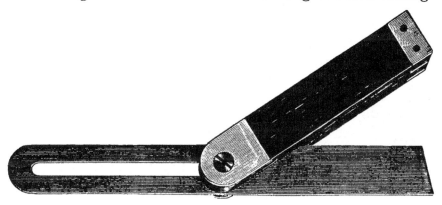

Fig. 33.—Improved Ebony Sliding Bevel.

appliance, with a movable blade, as in Fig. 33, so that it can be adjusted to any angle.

A *wooden square*, larger than the above-mentioned,

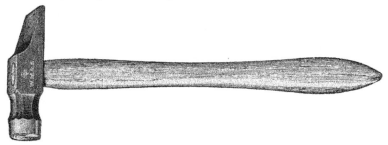

Fig. 34.—Hammer.

will be found very useful for panels and other large work. It can easily be made by the user.

Hammer.—The usual and most useful form of this for cabinet-makers is that shown in Fig. 34.

Mallet.—This is a kind of wooden hammer, and is useful where the iron one would injure either a tool handle or the work. It need not be heavy.

Punch.—This is merely a blunt piece of steel for driving the heads of hammered nails below the surface of the wood.

Compasses are used both for measuring and for scribing circles or parts of them. To prevent them opening or closing when in use an arm or wing is fixed to one leg, and projects through the other, in which is a screw by means of which they can be fixed at any opening. A pair of wing compasses is shown in Fig. 35.

Rule.—This, of course, is indispensable for measuring. It is made in a variety of patterns, and any but the very cheapest and commonest may be used. The latter are often unsatisfactory in the hinge or joint. A two-foot is as useful as any; it may be either twofold or fourfold, that is, with one or three joints. Both in regard to length and folds the choice depends entirely on the worker, some preferring one kind and some another. Indeed, a similar remark applies to all tools.

Fig. 35.—Wing Compasses.

Scraper. This is a thin piece of steel from four to six inches long, by rather less in width, and though apparently not commonly known among amateurs, is invariably part of the workshop outfit. As its name implies, it is used for scraping surfaces smooth, and may be regarded as an adjunct to the smoothing plane, or as superseding this when its use would be unsuitable, as in the case of veneered work. For ordinary purposes they are straight edged, but they are occasionally seen in various shapes for use on mouldings, &c.

Scraper sharpener.—This is a piece of steel for sharpening scraper edges, but many cabinet-makers prefer a 'currier's steel,' while many dispense with a special sharpener, and use the back of a gouge instead. Almost any rounded steel surface will answer the purpose.

Marking awl (Fig. 36) is merely a sharp-pointed piece of steel for marking with instead of a lead pencil. A useful form of a similar tool is that shown in Fig. 37, which has a chisel or knife end as well as a point. Anything similar will answer the purpose equally well. As

Fig. 36.—Marking awl.

Fig. 37.—Marking Chisel or Knife.

pencils are referred to in above sentence, it may be well to say that those generally used, instead of being round, are oval in shape, and are better adapted for the work than the ordinary kind.

The *oilstone*, or stone on which the edges of tools are kept keen, is a very important adjunct to the cabinet-maker's bench, and care should be used to get a good one. The qualities of the different kinds vary considerably, and even in each kind there is much variety. It is not always possible to determine, except by actual trial, the properties of any stone, and a good one is valued by its owner. Some stones are quick cutting, that is, they soon put an edge on a tool, while with others a good deal of rubbing is required to produce this. The principal kinds of stones used are Charnley Forest, Turkey, Washita (Ouachita), and Arkansas, as well as a few others, all of which have their admirers. For general good qualities I doubt if anything can surpass the Washita, which is also moderate in price. It cuts quickly, but, perhaps, does not give quite such a keen edge as the Charnley Forest and Turkey stones. The former is, however, slow cutting, and the latter is apt to wear irregularly. Good stones of any of the varieties named are to be got, and I cannot recommend any as being far superior to the others, nor, on the contrary, can I denounce any.

The stone can either be got loose or boxed, but if the former it should be cased by the user as soon as he can conveniently do so. The box generally consists merely of two pieces of wood, both hollowed out, one in which the stone is fixed and the other acting as a lid. Such a box, with stone, is shown in Fig. 38. To fix the stone in various cements are used, but nothing is better for the purpose than a mixture of hot glue and dry red lead.

Fig. 38.—Oilstone in Case.

To keep a stone in the best condition the oil should be wiped off whenever it is laid by, and never be allowed to harden on it. In course of time the stone gets worn down, and it becomes necessary to level it. This, which at best is a tedious job, may be most expeditiously done by rubbing it on a level board, liberally sprinkled over with emery powder. Sand may be used instead of emery, but it is not so quick. Another way is to use the grindstone, the sides of which, of course, are used for the purpose.

TOOLS.

Tools occasionally require grinding as well as sharpening, and the cabinet-maker may prefer to have this done for him instead of doing it himself, and may then dispense with a grindstone. Otherwise one will be necessary. As is no doubt well known, it is merely a circular slab of stone, fitted with a handle, and fixed in a more or less elaborate stand. If the latter has a trough to hold water and keep the stone constantly wet when it is turned, care should be taken to pour the water away when done with. To allow any portion of the stone to soak in water alters the density, and accordingly spoils it for grinding. A very large stone is not required, and the professional cabinet-maker may be told that the grindstone is not a personal tool or appliance, but is part of the workshop fittings.

Cork rubber.—This, for use with glass paper, has already been sufficiently referred to.

Hand-screws.—These are continually necessary to bind pieces of wood together till the glue has set in the joint. As will be seen from the illustration, Fig. 39, pressure is exerted by turning the two screws, the wood being held between the square jaws. They are made in various sizes, determined by the length of jaws. The nine-inch will probably be most useful to begin with, as there is comparatively little that cannot be done with them. If the screws work unpleasantly stiffly, a little soft soap and blacklead will be found a good lubricant. Oil should not be used, as both screws and jaws being of wood it would cause them to swell.

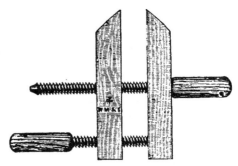

Fig. 39.—Hand-screws.

The *bench holdfast*, Fig. 40, is made of iron and is used to fasten down wood when being sawn or worked on in any way which requires it to be firmly held on the bench top. To use it the bench has a hole through which the straight part passes easily. The wood to be held is put under the lower end of the bent arm, and sufficient pressure is brought to bear by turning the screw.

Fig. 40.—Bench Holdfast.

Cramps are made both of iron and wood. The latter is generally a home-made article, and for most purposes is sufficient, but occasionally something more powerful is wanted, and the cabinet-maker is advised to have one, if not two, of iron. The general form is shown in Fig. 41, which represents a plain iron cramp. The wooden ones are similar in principle, the modifications being only such as required by the material. As the wooden

Fig. 41.—Iron Cramp.

cramps are seldom seen in tool-shops the cabinet-maker must either make them himself or get them made by a cabinet-maker or joiner. A good useful length is three feet and the cabinet-maker will rarely need anything larger. Instructions for making them will be found further on. The cramp is used for much the same purpose as the hand-screws when these are not large enough or cannot conveniently be applied.

Rasp and file.—These come in handy for a variety of purposes, which will suggest themselves when working. The most convenient form is that known as the half-round, and eight inches will be a suitable length.

Dowel plate.—As this has been already mentioned, it is only necessary to say that it can be obtained ready made, with several holes for different thicknesses of dowels. Bits, of course, will be required to match.

There are several other appliances which, if not absolutely necessary, are at any rate so useful, that they should be added to the outfit as opportunity occurs. Many of them being of wood may easily be made by the user, though, if he is a novice, he will find it difficult to do so in some instances. Should he not be able to manage by himself, he can easily get any experienced wood-worker to do what is necessary, and sufficient instructions will be found in the next chapter.

CHAPTER VII.

WOODEN APPLIANCES MADE BY THE USER.

Cramp — Extemporised Cramps — Shooting Boards — Mitre Shooting Boards — Mitre Block — Mitre Box — Square — Straight Edge — Winding Strips — Scratch or Router — Benches — Tool-Chest.

IN making the following tools, though their construction is neither difficult nor complicated, accuracy is certainly a necessity, for without it most of them would fail entirely in their object. On this account, they may be beyond the powers of the novice, who, as stated in the previous chapter, will find it advisable to get them made for him. If he is able to get any one to lend him things to copy, it may be an assistance, though without a few words of explanation he might not be able to construct them. I should say that he will find an even greater variety in the details of such appliances as form the subject of the present chapter than in the ordinary tools of the tool-shop. These, being mainly made in large quantities, are often met with of exactly the same pattern; but in the wooden appliances the maker can suit his own taste or fancy for some form in preference to others. Thus, while there is a certain resemblance—*i.e.*, the main features are preserved in each class of tool—there is a great difference in such details as size, kind of wood, and general construction. It is, therefore, more likely than not that any specimens the reader may have access to will differ in some respects from those mentioned here. He may, however, depend on these being good, workmanlike things. If he chooses to elaborate their construction, he can easily do so when he has gained the necessary experience. It is of course impossible to enumerate everything which may be made by the user, for it is by no means uncommon to find that an intelligent artisan has devised some appliance which he may find convenient for some special purpose. Whenever the novice thinks he can do the same he should not hesitate to do so, though let it be remembered it is no use encumbering oneself with things made from mere faddiness, and that it requires a good deal of experience to know whether a so-called improvement is of any advantage. Any one can devise altered forms, but only the experienced can develop improvements.

Cramp.—The bar of this may be any length, but a convenient one will be about 3 ft. The wood should be hard and strong, such as oak or ash, but it is by no means necessary that these only should be used. Stuff about 1 in. down by 3 ins. wide will be suitable, but these dimensions may vary considerably, as all that is wanted is that the cramp shall be strong enough. The lower edge is notched much after the style of an exaggerated saw, or it may simply have a series of half-round holes; anything, in fact, which will allow the iron pin connecting the sliding block to catch. The upper edge to within a few inches of the block in which the screw works has a groove ploughed along it. In this groove a corresponding tongue on the lower edge of the sliding block fits. The block is thus guided in a backwards and forwards movement, but is restrained from working laterally. The sliding block may be about 4 ins. high and perhaps a little wider on the base, and is best with the grain perpendicular. The strain is then not so apt to cause breakage as when the grain is parallel with it. Much, however, depends on the strength of the wood, and size also is of comparatively little importance. To form a connexion between the bar and block as well as a stop to the latter, an iron loop must be formed, the sides being of thin plate, and the stop and pin which runs through the sliding block being round pins, say, $\frac{1}{8}$ in. thick. These pins should be shouldered and then rivetted. The hole in the block should be large enough to let the one through it fit easily, and of course the rivetting can only be done after the pin is in. Block and bar should be of the same thickness. The exact shape of the block is a matter of fancy, and does not affect the working of the cramp.

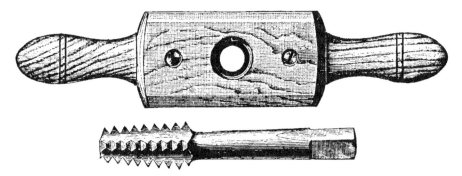

Fig. 42.—Boxes and Taps for Wood-screws.

The screw-block at the other end is a fixture, and it must be well fastened. Unless the maker has a box and tap for wood-screws (Fig. 42)—a somewhat expensive appliance and one not often seen in the cabinet shop—he cannot attempt to make the screws himself, but will probably be able to get a turner to do what is necessary. If he cannot, it may be suggested that a jaw and screw of a hand-screw can easily be adapted. To make a firm fastening, a hollow may be cut on each side of the bar of the width of the block. This has a slot cut in it to fit, and the two pieces may then be secured by two or three screws or wooden pegs. Care must be taken that the block is fitted so that the screw works parallel with the bar, and its height should be such that it is opposite the face of the sliding block. The point of the screw should have a small iron spike fitted into it. This may easily be managed by driving in a screw-nail, of which the head can be filed away and

Fig. 43.—Wooden Cramp.

a projecting point, say, $\frac{1}{8}$ in. long left. To prevent the wood splitting, a small ferrule of brass tubing like that on a chisel handle should be fastened on before driving in the nail. The end of the wooden screw may be cut to fit the tubing if necessary.

The completed cramp is shown in Fig. 43, and there will be no difficulty in recognising the various parts which are shown separately by Figs. 44 to 48.

Fig. 44.—Bar for Screw-block.

Fig. 45.—Screw-block to fit Bar.

Fig. 46.—Bar showing Groove.

Fig. 47.—Sliding-block showing Tongue.

Fig. 48.—Iron Link.

To use the cramp, the sliding-block is placed in about the position required, the loose link allowing it to be moved freely, but forming a stop when necessary. The wood to be cramped is then placed, and tightened up by turning the screw. To prevent the point injuring the work, place a piece of waste wood between them. Bruises from the moving block may be prevented in a similar manner.

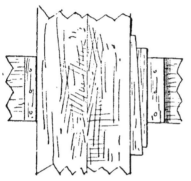

Fig. 49.—Improvised Cramp.

A rough and ready form of cramp may be made by simply nailing two pieces of wood across a board parallel with each other, and wider apart than the work to be cramped up. This is laid between them and tightened up by means of wedges. This makeshift cramp is shown in Fig. 49, and may come in handy when the regular ones are otherwise engaged.

A similar contrivance is that represented by Fig. 50, which consists merely of four pieces of wood of any convenient size nailed together. The work to be cramped is placed within the opening and tightened up as before with wedges. Other forms of cramping arrangements will no doubt suggest themselves if they become necessary.

Fig. 50.—Improvised Cramp.

Shooting Board.—This will be found of great assistance for truing up or shooting square ends and edges of wood with the trying plane, and it comes into constant employment. Its form will be seen in Fig. 51, and the construction is as simple as it looks. The wood of which it is made should be thoroughly dry, and of a kind not apt to twist or go. Good, clean pine will do very well. Three pieces will be required. The bottom one may be about 10 in. wide, and the upper one 3 in. or 4 in. less. The length must be determined by convenience, but about 2 ft. 6 in. will do very well. On the end of the upper board a cross-piece is fixed at exactly a right angle with the edge, beyond which it must not project. Inch stuff will do very well for all the pieces, the end one of which should be fastened with screws or nails. The wood which is to be shot is placed on the upper board, with one edge against the stop and the one to be planed just projecting over, so that the plane when laid on its side on the lower board can remove what is necessary. When end grain has to be trued up, it is done in the same way, the only difference being that the end of the board is presented to the plane. The trying plane is used in both cases, as its straight, flat side runs easily along the lower board. The edge of the upper one serves as a guide.

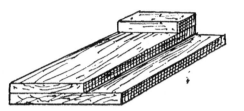

Fig. 51.—Shooting Board.

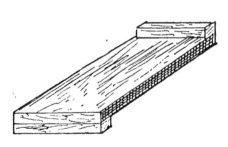

Fig. 52.—Bench Hook or Shooting Board.

A simpler and cruder form of shooting board is shown in Fig. 52. It consists of only the board, to which the stop is fastened as before, the bench top taking the place of the lower one. To keep the board in place on the bench, a block of wood may be fastened underneath to catch against the end of the bench top. Such a bench hook is also useful when planing surfaces, especially if the bench is a rough, uneven one, as to some extent it may be used instead of the ordinary bench stop. Of whatever form the shooting board is made, all the edges should be square.

The mitre shoot is for the same purpose, only instead of being for square corners it is adapted to those at 45°, or half a right angle. In case the novice does not know what a mitre is, he may be referred to an ordinary picture-frame, not one of the Oxford pattern. At the corners he will notice that the joint is a diagonal one, so that all the members of the mouldings meet uniformly. To ensure them doing so, the joint must be cut at exactly half a right angle, *i.e.*, for a square-cornered frame; and pieces whether of moulding or anything else joined thus are said to be mitred. Of course, if the frame or work is octagonal or other shape, the angle of the mitre varies accordingly, but it must be half of the whole angle. Thus for a mitred joint of 60°, each piece would be cut at an angle of 30°. The mitre, however, is more commonly required with square corners than with any other, so the ordinary mitre block is made accordingly.

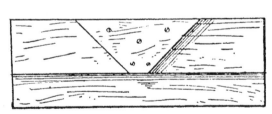

Fig. 53.—Mitre Shoot.

The lower boards are as before, the only difference being in the shape and position of the stop. After what has been said, it seems almost unnecessary to say that it must lie across the board at an angle of 45°, and as one guiding edge is obviously useless, the block is double, so that the end of a moulding can be shot either way. This necessitates the stop block being placed near the centre, as in Fig. 53. A simpler form of mitre-shoot, corresponding to the second shooting-board named, may be made from a piece of board with parallel edges, and a piece the edges of which are also parallel, placed across it as shown in Fig. 54.

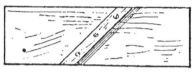

Fig. 54.—Simple Mitre Shoot.

The mitre-shoot, it must be understood, is only to be used for truing up mitres after they have been cut. To guide the saw while this is being done, a mitre-block or a mitre-box is required. The former is the simpler of the two, and consists of two pieces of wood, on one of which, or rather in the angle formed by the two, the wood to be mitred is placed, while the saw is guided in

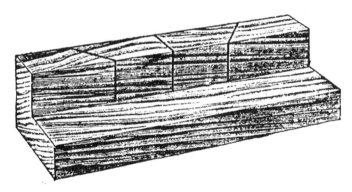

Fig. 55.—Mitre Block.

one or other of the vertical cuts, which, of course, are at the proper angle across. The centre cut observable in the illustration (Fig. 55) is straight, and is for sawing square ends on small pieces of wood. A useful size for the block is about 1 ft. 6 ins. long, the lower piece being about 6 ins. wide, and the upper one 3 ins. This latter should be of a good thickness, say not less than 1½ ins., for the wider it is the better the guidance of the saw. The novice will please note that the cuts must be perfectly perpendicular to the bed of the block.

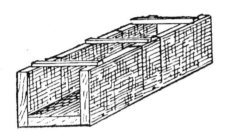

Fig. 56.—Mitre Box.

The mitre-box, as represented in Fig. 56, is preferred by many, and has the advantage of guiding the saw on both sides of the stuff being operated on instead of only on one. For this reason it will no doubt be considered preferable, by the novice at any rate. It may be about the same length as the other. I may say that the size of the wood being worked is almost the sole guide to the dimensions of any of these wooden appliances; and, where figures are given, they are only to be taken as giving some idea of what is generally suitable. Provided the corners are square inside, the cuts vertical and at the proper angle, little else is required. A roughly made box, if these details are right, is quite as efficient as one put together by means of the finest work. Pine is suitable stuff. Width of each piece may be about 4 inches. Screws will fasten them together, or even ordinary nails. The cross-pieces on top are merely for the purpose of affording support to the others, and may often be omitted without risk if the box is made of hard stuff. A straight across cut may be made in this as in the block. With either box or block the tenon or dovetail saw is used, and not the larger kind.

A *wooden square*, referred to in a previous chapter, is

WOODEN APPLIANCES MADE BY THE USER. 115

among the things which may be made by the user. Its shape is identical with the small square already shown in Fig. 31 (page 98). The long, thin blade of it is replaced with one of wood fitted into a thicker piece, as in the illustration referred to, or simply nailed on to it. This, though perhaps of simpler construction, is not so good a form as the other. The essentials, of course, are perfectly straight edges and square angles. Dry, hard wood should be selected, and one of an even grain which will not be apt to twist. The thin blade may be of $\frac{1}{4}$ in. stuff, $2\frac{1}{2}$ ins. wide, and 18 ins. long, the other piece being less, but about 1 in. thick. The thin piece will be fastened in a mortise in the other, and be secured by a few screws. The truth of the square can be tested by laying it on a piece of wood, as shown in Fig. 57, drawing a pencil line along its thin edge, and then reversing as represented by dotted lines, and drawing another line either on or close to the other. If the two ruled lines

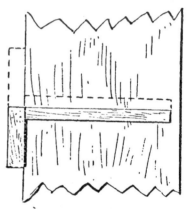

Fig. 57.—Testing Square.

coincide and are perfectly parallel, the square is all right. If, however, there is any divergence of the lines, it will be an easy matter to plane the edge of the square till the error is rectified. To reduce any risk of the square getting out of truth, it will be as well either to polish or to oil it. It can of course be tested from time to time if considered advisable.

Straight edges are long pieces of wood for ruling or testing purposes, with, as their name indicates, at least one straight edge. This must be really straight, and not merely something near. Any thickness of stuff may be used, but there is no occasion to have it more than of ¼ in. or ½ in. The edge may be tested in a very similar manner to that directed for the square, viz., by drawing a line with a pencil against the edge supposed to be straight, then reversing the piece, and drawing another line. When doing this it will be well to reverse the ends of the straight edge, and not merely to turn the wood over, as any error will then be more readily detected.

Winding strips, the use of which will be found explained in due course, are merely two straight pieces of any kind of wood of exactly equal width, which may be about 2 in. or less, and sufficiently thick to stand on edge. It is sometimes considered better to have the wood bevelled or thinned away to the upper edge, but this shaping is a matter of taste, and can hardly be considered necessary. It is, however, of the utmost importance that the width should be equal in both and uniform, or the strips will be worthless. The length may be from 1 ft. 6 ins. to 2 ft., for the longer they are in moderation the more easily will the errors they are intended to detect be discovered.

The *scratch or router* is an exceedingly useful tool for working the small beads and hollows now so much seen on furniture, and is also available for forming small mouldings, chamfering, and a variety of purposes. Although there are many improved forms, or what are said to be such, of beading routers, the rough, home-made thing may be as serviceable as any, and has advantages which they do not all possess. For one thing, it is cheaper than any, for its cost is merely nominal, while the difficulty of making is very slight, so that it would be inconsiderate to those who wish to fit themselves out economically not to refer to it even if others prefer the more ornamental tool supplied by dealers.

The complete tool is represented in Fig. 58, and consists of two pieces of wood, shaped as seen, fastened together and holding between them a small steel cutter, or rather scraper, with its bottom edge filed to the desired form of bead or hollow. The trouble of making lies almost entirely in these cutters, for on them depend the shape and appearance of the beads, &c., formed by them.

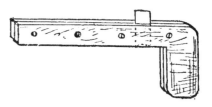

Fig. 58.—Scratch or Router.

The stock is formed of two pieces of any kind of wood, though hard is better than soft. They may be of ½ in. stuff, and say about 9 ins. long. The wide or handle end is cut out square with the narrower position, and works against the edge of the wood like the block of a gauge, to the action of which the scratch is very similar. The lower edge of the two pieces is best when rounded off slightly on the part which works on the wood being beaded. The narrow portion may be about ¾ in. wide, and the other anything in reason. Screw-holes are bored through at intervals to receive a couple or more screws, between which the cutter is held at any place required. By screwing up tightly the cutter is firmly fixed, and can be removed for the substitution of another by doing the reverse.

The cutters themselves require more attention. Any number or variety may be made as wanted, but as the action is a scraping one on the wood they should be kept of moderate dimensions, as to use them over ¼ in. or ⅜ in. wide would entail a considerable amount of labour. Although each cutter should be narrow, bands of beadings can be made of any reasonable width without much difficulty

Fig. 59.—Cutter for Bead.

by altering the position of the blade. Thus with a cutter shaped as in Fig. 59, a single bead or reed like that shown on Fig. 60 would be produced, and by altering the position of the iron several beads may be worked close together, as in Fig. 61, while by sub-

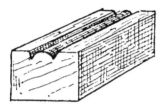

Fig. 60.—Bead.

Fig. 61.—Rows of Beads.

stituting another in place of the bead a hollow may be worked, as in Fig. 62. The shape of the cutter to form this hollow is of course round, as in Fig. 63. A large hollow may be worked with a cutter shaped as in Fig. 64. Only half of the hollow is formed with it at once,

Fig. 62.—Beads with Hollow between.

Figs 63 & 64.—Cutters for Hollows or Flutes.

but to make the other it is only necessary to reverse the iron. If there is any irregularity afterwards visible in the centre of the hollow through the cuts not exactly matching, a rubbing with glass paper, held in this instance over the edge of a piece of wood rounded to match the hollow, will remove it. Scratched beads generally want cleaning up with glass paper, and care must be taken not to rub them away too much. It should always in such cases be used over the edges of wood, either cut quite thin or to correspond with the beads, &c. By having several irons, and altering their combinations, an almost endless variety of beadings and members can be worked. For chamfered edges a plain edge cut on the skew is all that is necessary, it being placed close in the corner of the stock as in Fig. 65, or as in Fig. 66.

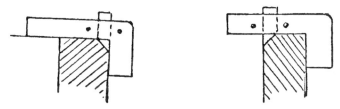

Figs. 65 and 66.—Scratch for Chamfering.

To work mouldings—only small ones, or rather small members of mouldings, for a large moulding may be entirely composed of small members—the cutting irons may project to any required depth instead of being close up, as when scratching plain surfaces. Speaking generally, the cutters may be fixed in the position deemed most convenient to accomplish the work intended.

As material from which to form the cutters, nothing is more suitable than pieces of broken band-saw, but in case the reader may not know anything about these, let it be said that metal about the thickness of a scraper is suitable. It must not be so thin and weak as to bend, but nothing is gained by having it inordinately thick.

To shape the ends of the cutters, a few fine files will be needed. The cutters are filed straight across, and not sharpened like an ordinary cutting blade. At the same time the edges must not be round, for to make them so would render them almost useless. To lessen the labour of filing, the steel should be held in a vice. It may convey a useful hint to say that the needed shape can be filed out with more confidence by fastening the steel between two pieces of thin wood, the ends of which have been shaped to that wanted on the cutter. Some are in the habit of sharpening the cutter much as if it were a scraper, while others smooth away any roughness with oil-stone slips. Either of these courses is troublesome, so that many content themselves with just using the file. Those who are accustomed to work in steel will no doubt think fit to prepare the blades before filing, and then harden and temper them afterwards. Such treatment, however, is not necessary, so that no directions about it are called for here.

Many other little odds and ends in the shape of wooden appliances are occasionally seen, but the principal, and those which are almost invariably found in use for general purposes, have been named. The others are more or less fancy tools, or are only used for special purposes. Any such, like the veneering hammer, will be found described elsewhere in connexion with the work to which they belong or which may be facilitated by using them.

It has been a matter of consideration with me whether to include the bench among the things which should be described, but I am inclined to think a detailed description is unnecessary. The professional cabinet-maker is seldom asked for an opinion on what kind he would like. He just takes them as he finds them, so that directions for making benches are not necessary for him, especially as by keeping his eyes open he will see more varieties than could possibly be described here. The amateur would find bench-making very uninteresting work, little more indeed than heavy joinery being wanted ; and, moreover, if he is competent to make it, would probably prefer to embody his own

WOODEN APPLIANCES MADE BY THE USER. 121

ideas in it. If he cannot make it he can either buy one ready or have it made for him by any carpenter. Bench and bench-making is an almost inexhaustible subject, and one which would require a book to itself to treat it fully.

The principal features about a bench should be firmness and solidity : so long as these are secured the rest is merely a matter of detail. No bench, however, could be considered complete without a bench-screw or wooden

Fig. 67.—Cheap Wooden Bench.

vice and a stop to prevent wood when being planed from slipping off. A good useful size for cabinet-making is about 6 ft. long by 1 ft. 6 ins. wide by 2 ft. 6 ins. high, and for most purposes one considerably less on top would do equally well. The amateur, therefore, who may be cramped for room must not fancy he cannot indulge in cabinet-work because he has not got a full-sized bench. For his accommodation many tool-dealers keep small benches, one of which, measuring 34 ins. long by 26 ins. high by 13 ins., is sold with a variety of tools for 22*s*. I may, however, say that I have not seen this bench, which is probably more suitable for a boy than for a man. A fair-sized bench, 4 ft. 6 ins. long by 1 ft. 3 ins. wide, as shown in Fig. 67, is obtainable for 22*s*. 6*d*., and it is really a useful thing, though rather too light for the rough wear and tear of a trade workshop.

For this the bench made by the Britannia Company, and represented on Fig. 68, is admirably adapted, the

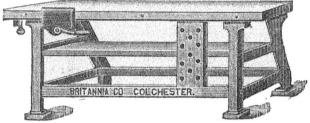

Fig. 68.—Iron Standard Bench.

supports being of iron, held together by bolts and screws, so that it can easily be taken to pieces when required. The top, of course, is wood. It is made in various sizes from 5 ft. by 1 ft. 6 ins. to 7 ft. by 1 ft. 10 ins., and is fitted with a patent 'Instantaneous grip vice' in lieu of the more ordinary wood-screws and blocks.

The 'German' Bench, of which a modified form and one better adapted to English usage than the original is shown on Fig. 69, is very highly spoken of by those who are accustomed to use them. It will be seen that the principal difference from the ordinary bench consists in an end block and screws in addition to the usual one in front. By means of it work can be held at both ends between the pegs, one of which can be placed in the most convenient position in a series of holes near the front edge. The end screw and pegs can also be used for cramping up door frames, &c.

The bench is illustrated as sold by various dealers, but I consider it would be much improved by the addition of a perforating sliding board between the top and the lower rail very similar to that seen in the before-named iron framed bench. The board, it may be said, is useful for supporting long pieces, one end of which is held by the screw, as a peg can be put in any of the holes. The board being movable can be altered to the most convenient position. It can easily be added by any one, as it is merely a piece of wood of sufficient substance, say, 1 in., with top and bottom grooved or tongued to fit corresponding tongues or grooves on top of the rail and below the top. The construction of the bench is so clearly shown that no one who can manage the manipulative work will have any difficulty in making or adapting an existing

Fig. 69.—German Bench.

one by adding the essential feature, viz., the end screw and row of holes.

It is the less necessary to say anything much about the construction of benches, as those used in many workshops are very rough old things, so crude and coarse that the amateur who may be accustomed to seeing them in their new condition might think them no good. Still much of the best furniture is made on such as these, for a clean smooth top is by no means necessary to those who can work, though it must not be winding. A good bench will no more render the novice a competent worker than the possession of the newest thing out in the way of tools. A smooth top may be desirable, but even if it is worn and battered to such an extent that it does not answer for all kinds of work, the remedy can easily be found in a board temporarily placed under the work in hand. Whatever care is taken, a bench top is bound to show signs of wear in course of time, and if these are from legitimate use no exception need be taken to them. It is of far more importance to keep it free from nails or screws, though these may be driven in temporarily when required to hold work.

The bench is for working on, not for ornament. The top should be kept as clear as convenient from odds and ends when using it, and as many tools are constantly being required it is a usual thing to have a trough or tray along the back, so that small tools can lie in it and be out of the way of the work.

A chest of some kind will be useful to keep the tools in; indeed, for the professional worker, it will almost be a necessity for more reasons than one. Very little need be said about it except that it should be strong, and so arranged that the contents can be kept in an orderly manner and any tool be got at easily.

A very convenient form, and one frequently adopted, is fitted inside with shallow drawers or trays less than half the width of the box. In these the smaller tools, chisels, gouges, bits, &c., can be placed, the rest of the space being occupied with the larger things. The trays may be laid on small strips of wood nailed to the ends of the box, or they may be more highly finished as drawers. There is no recognised pattern of universal adoption.

I cannot speak highly of the combined benches and tool-chests which are sometimes seen in the amateur workroom. They may be convenient in some rare cases; but, as a rule, they, or at any rate the bench portions, are more suitable for odd jobs than for real work, unless they are so large that as chests they are awkward in case of removal. A good bench can easily be made portable, so that the top is the largest portion which has to be moved, and a separate tool-chest need not be unwieldy or unsightly.

Speaking generally, the same rule applies to combination tools as to combined benches and tool-chests, *i.e.*, they are not to be recommended indiscriminately. Many of them are very ingenious, and it may also be added, costly, but it is rarely that a combination is equally good in all its forms. In one or other it may be as serviceable as a separate tool would be; but, as a rule, it will be better to get simple forms, and not be taken with ingenious combinations.

With regard to new forms of tools, or so-called improvements, the amateur need be in no hurry to get them. Let him wait till they meet with general approval in practical workshops. They will be adopted quickly enough if they are really improvements by professional workers. Many of these improvements have, no doubt, been devised by them, but for all that cannot be regarded as of much practical utility, but more often as fads and fancies which have been developed and which some enterprising tool-maker may have taken up. It must be remembered that tool-makers and dealers are not generally practical cabinet-makers, and that these latter will be safer guides as to the best forms of tools. It is astonishing how quickly any really advantageous improvement is adopted, and if any purporting to be so remains unused for any length of time among practical cabinet-makers, the amateur may depend there is not much in it. This may be 'straight talk,' which will not be altogether regarded with favour by some; and, of course, if the reader likes to spend his money over fads there is nothing to prevent him doing so. The professional worker, though it is to his interest to turn out his work quickly and well, does not, as a rule, care to waste money on things which his experience shows him are not improvements in reality though they may be called so. Were it expedient to do so, I could name many of the class of things referred to, which show pretty conclusively that amateur workers are numerous, and that they go in largely for improved tools or it would not pay dealers to keep them, as they are never seen in practical cabinet-making workshops.

CHAPTER VIII.

GRINDING AND SHARPENING TOOLS.

Angles of Cutting Edges — Workshop Practice — Grinding and Sharpening Edge Tools — How to Hold Them — Sharpening Scrapers.

THE importance—indeed, I may say the absolute necessity—of keeping tools in good order has already been referred to, but it would be little use to tell the novice this unless he were shown how to get them in condition for work and keep them so. Half the battle, no doubt, is to get them right at first, for they are kept so with comparative ease then. To attempt good work without properly sharpened tools is quite out of the question, and there are, I imagine, few readers who will be unable to get a cabinet-maker or joiner to show him a few tools sharpened just to give him a start.

I say this, as many amateurs seem unduly concerned to know all about the correct scientific rules which govern the sharpening of tools, the proper angle, and so on. Well, let me tell such that in the practical workshop these matters are not regarded. No one ever thinks of measuring the angle; he knows about what suit, and that is sufficient. If, then, the reader will look at Fig. 7, on page 84, he will be equally well-informed, and he need not bother about measuring angles. The cabinet-maker is working in wood, be it remembered by those who do not care about 'rule of thumb' work, and, for all practical purposes, an edge that will do for one kind will do for another. If I may say so, without any one supposing that blunt tools will do at any time, the softer the wood the sharper the tool should be to ensure a clean cut. This, though, is more theory than anything else, for a chisel which will cut oak cleanly and perfectly will do the same with pine. To reduce theory to practice, which, after all, is the main thing, at present at any rate, a cutting edge may be tried across the grain of a piece of pine or other very soft wood. If it makes a clean cut, not a tear or a bruised-looking one, it will do for any wood, and may be considered quite satisfactory. This hint, to those for whom I am specially writing, will be in practice worth a volume of theory.

In many, perhaps the majority of cases, the purchaser of new tools will be able to have them delivered properly ground. If not, and he does not possess a grindstone of his own, he will have no difficulty in finding a carpenter or some one who has one. It may be said that any tool-shop will undertake to grind tools, but it would hardly be fair to get this done locally if the things are bought elsewhere—that is when they are new. Regrinding is a different matter somewhat, and tools, properly used, do not often require this. Of course, if their edges get notched, they must be reground before they can be sharpened. Grinding, it may be explained, only partially sharpens the tools, the cutting edge being afterwards given with the oilstone. It might be done with the latter only, but the labour would be greater than necessary.

Grinding is somewhat dirty work, as the stone has to be kept wet with water, and also laborious. If some one can be got to turn the stone it will be easier, besides leaving both hands free to hold the tool, which must be kept steadily in the right position, for a clean bevel is wanted, and not an edge shaped like Fig. 70. To get a straight edge across the blade is not

Fig. 70.—Badly Ground Edge.

less important, for if this is ground away more on one side than the other, or rounded off towards the corners it will be defective for most tools. The aim should be to get a straight and sharp edge as represented in side and front view by Fig. 71. The stone may be turned either to or from the edge being ground, but the former is the better way. When turned from the tool there is considerably more risk of a 'wire' edge, a kind of loose burr, being formed. The grinding should be done only on one side of the tool, and not on the flat as well. With a little care no great difficulty will be experienced in grinding properly. Tool holders have been devised for keeping the edge evenly on the stone.

Fig. 71.—Properly Ground Edge.

The oilstone has been referred to, so nothing more need be said about it. To use it properly, that is, to put a keen edge on, requires a little knack, and beginners often seem to experience some trouble in sharpening, though there is really very little difficulty in doing it. The stone being lubricated with oil the tool should be rubbed firmly backwards and forwards, not jerkily or anyhow. The handle or end of the tool may be held in the right hand, and the left presses on the blade. At first the pressure on the stone may be fairly great, but as the edge becomes sharper it should be decreased till, for the last rub or two, it is almost nothing. The tool should be kept as kept as nearly as possible with the ground bevel flat on the stone, just raised enough to ensure a sharp edge being formed. If held more upright it would not at first much matter, but regrinding becomes necessary sooner than it otherwise would. To remove any wire edge—indeed, in any case—after sharpening as directed, lay the other side of the chisel or tool on the stone, on which it should be flat or nearly

K

so, and give it a rub or two in that position. Of course, the same general precautions must be observed as when grinding, and with a very moderate amount of practice the novice should be proficient in sharpening his tools. Saw sharpening need not be further referred to, as it was sufficiently dealt with elsewhere.

The scraper, to those who are not acquainted with its powers, may seem an insignificant tool, and hardly worth attention in sharpening. The fact is though, that unless this is done properly, the scraper is practically worthless, for it simply rubs instead of scrapes. As the edges are alike they may all be sharpened, but usually only one of the long ones is. If the steel is not in condition when got, the edge, assuming it to be for flat work, must be made straight either on the grindstone or the oilstone. To make it perfectly square it must be rubbed on the oilstone while being held upright. Now comes the important part of the sharpening. One end of the steel resting on the bench top, the other is held by the left hand, with the edge to be sharpened towards the worker's right. With this one draw the scraper sharpener smartly upwards, firmly pressing it against the edge, two or three times. Reverse the ends of the scraper and repeat. Mind that the sharpener is held square across the edge of the scraper, or, instead of being sharpened, with a slight tendency to burr, it will be round. To make all clear, the edge should have sharp angles, as represented in Fig. 72, and not round, like Fig. 73. A properly sharpened scraper will remove shavings in a manner which might astonish the novice who has never seen one used.

Figs. 72 and 73.—Scraper Edge, sharp; Ditto, round.

CHAPTER IX.

GENERAL DIRECTIONS ON THE USE OF TOOLS.

Sawing—Planing—Scraping—Boring with Brace and Bits—Use of Winding Sticks—Circular Saw—Lathes—Fret Machine.

THE tools being ready, it may fairly be supposed that the novice is anxious to use them, and to enable him to do so general directions will be given. It must, however, be premised that the following remarks are not to be taken as if the work mentioned were so many exercises which must be worked out or practised before anything further can be done. It will, no doubt, be a means of acquiring experience if the novice does actually follow out the instructions as they are recorded, but they are intended to be of wider application than this. They are the points to be put in force when actual work is being done, and any trials which may be made with loose timber, that is, pieces which are not intended to be formed into anything, will be more by way of experiment than anything else. The novice must not think that he will be able even by following the clearest directions in the most careful manner to use the tools properly at first. He must practise before he can gain facility, although there are some who seem to think that if they are told or shown how to do a thing they ought to be able to manage equally well. The things look so very easy when done by skilful hands that beginners sometimes apparently forget that it has taken time for even the cleverest worker to learn. The young practical mechanic, of course, learns almost insensibly how to hold and use his tools, for if he goes wrong he is checked at once by those over him. The amateur, however, must rely more on himself, and it is principally to aid him that this chapter is written. I may remind him that though directions for holding and working the tools are given, he need not follow them slavishly, if, after practising them, he finds that some other method suits him better. At first, though, he should follow them as closely as possible, even though the movements may seem awkward, as they undoubtedly will. Their use may be compared to learning to write. At first we are taught to hold a pen in what to the child seems a most constrained position, afterwards as practice emboldens we hold our pens to suit ourselves. Just so with tools, for every worker develops a style of his own, and ceases to be guided by hard and fast lines. Of course, only the most elementary usage of tools can be treated of here, for as he becomes more familiar with them the learner will see what is best to be done in any special case requiring different treatment. His experience will, as it were, insensibly widen.

As wood has to be sawn to sizes, or, as it is said, the stuff is got out, before a job can be begun, sawing may first be attended to. Let us suppose it is a plank, not small bits, to be worked on. The first thing will be to mark the lines which are to serve as guides. Needless to say that these must be straight. The board may either be supported on sawing trestles or on the bench top, the latter being the manner common among cabinet-makers. Those who have become accustomed to the other need not discard it, though it will be of advantage to saw with the latter. The trestles are merely supports for the wood, and may be of any convenient height.

When commencing a cut, place the saw teeth against the line, the handle being in the right hand. To steady the saw place the fingers of the other hand near its edge, and to give it a good start draw the saw upwards. Doing this a slight cut will be made, and the jumping about of the saw from the line on the first down stroke of the saw, so often seen when amateurs begin sawing, will be avoided, and the sharp edge of the wood near the saw kerf will be uninjured. It is presumed that ripping, that is, sawing with the grain, is being done. The sawyer should work with a steady action, remembering that the actual cutting is done on the downward stroke. The precise angle at which the saw is held is not of great consequence, and should be that at which the sawyer has the most command over his work. If the handle is too low down, that is, with the saw edge not sloping enough, the cutting in the wood between its top and bottom surface will be so long that the labour might be considerably more than it need be; very similar to that which would be exerted by cutting through much thicker stuff. Let any one try the extreme in this direction, viz., to saw with the edge all along on the wood, then raise the handle somewhat, and the difference will soon be noticed. At a certain angle or near about, every one will find he can saw better than another, and that is the right one for him. The thrusts should be as straight as they well can be, always in one line without swaying the handle backwards and forwards, and not altering the angle of the saw more than can be helped. Work both from the shoulder and the elbow. At first the saw will evince a decided tendency to wander from the line, and this must be carefully guarded against, for the slightest deviation at the beginning may at the end of a long cut have become a very serious one. With an irregularly set saw it is impossible to cut straightly. The edge will stray from the line.

If the saw is held with its edge perpendicularly or too nearly so the action will be more constrained than it need be, and there would be greater difficulty in keeping to the line.

Now, in addition to sawing straightly along the line, it will be remarked that the other direction of the blade, a sideways one, must be carefully watched, for if the saw slopes to either side the sawn edges cannot be square. If anything the tendency with the beginner will be for the saw to lean over to him as the cut extends, even if it was right at first. This must be guarded against, for when once a saw cut gets wrong it is not altogether easy to get the line right again. Prevention is better than cure, and frequent recourse should be had to the square. Put the block of this on the wood with the steel blade upwards and its edge against the side of the saw. If the two touch equally the work is all right; but, on the contrary, if the saw touches only the top or bottom of the steel, and, of course, gradually widens the distance between to the other end, the saw is slanting in one direction or the other. It seems almost needless to say that the sawyer stands with the board a little on his right, and moves backward as the cut progresses.

Cross-cutting is done in very much the same way, but it may be well to note that the wood when nearly sawn through must be held up near the saw, otherwise it will probably drop, and in doing so break off a splinter from the edge of the other piece. The splinter, of course, can easily be sawn off, but then the break in the other cannot be so easily repaired.

To rip with the wood on the bench top the screw holdfast (p. 104) will be found very useful, indeed indispensable, unless some one can be got to hold the wood, or it is secured by hand-screws, firmly. If the board is a long one it is best not to let the end project too far from the bench, so that the saw may work fairly close to the top. If it projects too far the board yields too much, bending under the pressure of the saw. This, and the resulting spring upwards when the saw is brought up, make the work awkward. It is an easier matter to alter the position of the board as the sawing progresses.

When using trestles the edge of the saw is generally towards the worker, but in the way now being described the reverse is the case. The sawyer follows the saw, the handle of which is held in both hands. But very likely some readers who have only seen the other method adopted may be inclined to think that this is not the proper way. To such I can only say that any way which is practised by skilful craftsmen is 'proper,' even though it may be unknown to the general public. That good and clever makers do make use of this method very largely when ripping is well known to every one who knows anything about cabinet-making, though strange to say I do not think I have ever seen mention of it in any publication. Those who prefer it, claim, and not without reason, that it has many advantages over that which is regarded as the ordinary way; it is, therefore, entitled to a trial. If a saw binds or works very stiffly, the sides may be slightly greased with advantage. When using the smaller saws, the procedure is much the same, limited by the backs, which of course will not allow one of them to make a long cut through. The bow saw is worked perpendicularly.

Planing is of more importance than sawing, so far as the actual finish of the work is concerned, and will probably present more difficulties to the learner. These are reduced as much as possible by the length of the planes first used, allowing the blades first to remove those portions which are highest, and gradually working down to a level surface.

The jack plane is first used if the wood is very rough and uneven, but in much of the machine-planed stuff which is often met with there is little occasion for it, and the planing may be begun with the trying or even with the smoothing plane. Everything depends on the state of the wood, and there is no better guide to the plane to be used first than the judgment of the worker. The jack-plane iron may project as far as it can consistently with easy working; but if too far, instead of shavings being cut pleasantly the tool will chatter and stick. To set the iron properly, the plane is turned over and looked along the bottom from the front end, so that the extent to which the edge projects can easily be seen. Perhaps the best way for the novice who has no experience will be to fasten the iron so that its edge is just within the mouth, when of course it will not act. Then tap it on top till it projects to the smallest extent and try it on the wood; it will probably then plane a little, but not sufficiently, so a few more taps should be given to the iron till satisfactory shavings are removed. The worker will easily be able to tell when the iron projects too much, and careful observation will be worth more than a bookful of directions. It will be well to note that should the iron project more on one side than on the other—that is, not lie evenly across the sole—it may be rectified by tapping it on the side edge near the top. The setting may also be modified by tapping the back or front end of the plane if more or less edge is required. Tapping the back gives more, while less may be got by doing the same at the other end. Hard blows should not be given, and in a very short time the novice will be able to set his plane irons by this means. When planing a long board, it may be found at first that the tendency is to take more off one part than another, and it will probably be at the end of the board near the stop. This and any other bad habits must of course be watched against, as it is of course necessary that the wood should be levelled equally, and kept at the same thickness throughout. It may help the beginner to suggest that he may almost try to plane hollow at first. This with a long plane he will find almost impossible, but the endeavour will counteract the tendency in the other direction. Of course, if a man tries deliberately to plane hollow he may do so to some extent, but it is by no means intended that he should do this.

The jack plane will somewhat level the wood, but the marks of the iron will be clearly visible, and may be removed with the trying plane, which will be finer set than the other. After the trying plane the smoothing plane is used when necessary, and it should have so little edge that only fine shavings are removed. If not carefully worked it may be well to point out that owing to its short length it is quite possible to plane hollow, especially towards the centre of a long piece of wood, or otherwise spoil the level surface. The truth of the surface can be tested from time to time by means of a straight-edge. Lay this on the board and notice whether it touches equally wherever it is placed—across, lengthwise, or diagonally. By having the straight-edge between the worker and the light a very small error may easily be detected. Try to look under the edge, along the board as it were, and the light showing through will at once proclaim any inequality. I don't know whether it may be necessary to say that as the object of planing, unless with the express purpose of reducing thickness or width, is merely to level and smooth the wood, and not to make shavings, any work beyond what will do so is superfluous.

When planing wood on the surface, the board is placed against the bench-stop, and the plane is worked in the direction of the grain as much as possible. This latter remark also applies to edge planing, but when this is being done it is generally more convenient to hold the wood by the bench-screw than in any other way.

With shooting boards the plane is placed on its side, the shooting-board lying flat on the bench top. One hand holds the wood being shot in position, the other works the plane.

When planing in the ordinary way the right hand holds the handle of a jack or trying plane, or the block just behind the iron with a smoother. The other hand holds and presses down the block in front of the mouth— fingers on one side, thumb on the other. The smoothing plane, if both hands are used with it, which is not always necessary, is from its small size held with the hand over the top front edge rather than over the top only. When drawing a large plane back do not press it on the wood, but ease the pressure and rather turn it a little over, so that more of the edge furthest away than of the whole sole is on the wood. The smoothing plane is so light that it almost naturally is lifted or eased from the wood when it is being drawn back.

If an edge is being planed, the trying plane or jointer, which is merely a long trying plane, is used, and is held in a somewhat different manner from that already described, at any rate so far as the left hand is concerned. The thumb is now placed on top with the fingers below, so that the finger-tips act as a kind of guide to keep the plane in position during the thrust. The smoothing plane should not be used on long straight edges, as it is too short.

Iron-soled planes run more easily when a little grease is used on them, so that some should always be handy. A piece of bacon rind does as well as anything, either for them or saws.

From the foregoing directions the way in which to use any other plane may be gathered, as in principle they are all alike.

As the scraper may be considered an adjunct to the plane the way to work it may be described here, and it should be remarked that its use is greater on hard wood than on soft. On the latter it is indeed seldom used, as it is not necessary and would only be a waste of time. The wood, as in planing, is placed against the bench-stop. Though a small tool both hands are used to work it. It is held with thumbs on one surface and the fingers on the other. Now, supposing the plate to be upright, slope the top edge forward so that the sharp angle of the bottom is in the best position for scraping. This is done by pushing the scraper forward and using considerable pressure. To hold the scraper in one hand and simply scrape about anyhow, as I have seen some amateurs do, is no use whatever. With a proper edge and proper use real good thin shavings are made. To scrape a large surface is hard work, and especially tiring to the thumbs. Though apparently simple, some knack is required to do the work thoroughly.

The use of the brace and bits is so obvious that it is only necessary to say that too great a pressure should not be exerted or the bits may be bent. The thinner and lighter the bit the less it will bear, but it is seldom anything is gained by using excessive pressure. One hand turns the brace while the other presses on the knob on top, though occasionally it is more convenient to place this against the worker's chest. In this position the base is kept more steady while being turned. At first a difficulty will be experienced in boring straightly, *i.e.*, perpendicularly to the surface, but by a little practice this may soon be surmounted.

When boring right through a piece of wood with a centre-bit, bore from both sides in order to ensure clean edges. Bore first till the pin comes through on the opposite side. This gives the exact position in which to place the bit when boring through from the reverse. If the wood is very thin place a waste piece underneath it, and in any case the pressure for the last few turns should be light to prevent the clearing or bent edge tearing into the wood.

The use of the winding sticks may not be clear to the novice, and as they are of considerable importance, especially when planing, and may be used for other purposes, such as testing door frames, though an accurate worker hardly needs them for the latter, it may be well to give an explanation about them here. It will be noticed that some boards are twisted, not merely hollow or rounded across their width, but lengthwise. Lay such a board on a bench top or other flat surface and it will not lie level, one corner will be up. Such a board is said to be 'in winding.' To make the meaning of 'in winding' clear, as it does not simply mean that a board is bent equally in any direction, for it may rather be explained as part of a twist or screw given to the wood, let the novice take a board sufficiently long and thin to be slightly twistable. Two people now take hold of it at opposite ends and facing each other. Now let them try and twist the board by depressing it and raising it at opposite corners, when it will, though of course only temporarily, be 'in winding.'

To detect a slight winding is not always easy without some aid. This is got with the sticks. Place one across near one end of the board and one near the other. Now, on standing at either end and looking over the nearest strip, it will be very easy to see whether the two edges coincide or not. If they do not, but instead one appears higher than the other or sloping in different directions, the board is 'in winding.' Figs. 74 and 75 will make the position of the sticks clear. In the first, they are shown on the board looked at from above; in the second, as looked at from the end, showing that there is a twist in the wood.

Any winding of this sort must be removed by planing, and it will not be altogether easy for the novice to get such a board into workable order. It may be as

GENERAL DIRECTIONS ON THE USE OF TOOLS. 141

well to say that a very much twisted board should be discarded when possible.

Before leaving the subject of tools, it may be

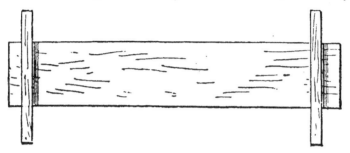

Fig. 74.—Winding Sticks on Board.

perhaps advisable to explain why no mention has been made of machinery for the cabinet-maker. By the use of machinery a good deal of manual labour is saved, but it is not adapted for use by the amateur or small cabinet-maker. Power of some kind is required to drive it, and only a large output could make it answer. Steam or power machinery is beyond the intention of this work altogether, for though useful in its way it would be of no use referring to it for general furniture making,

Fig. 75.—Winding Sticks showing Board in Winding.

It is, however, quite conceivable that many who have neither a steam nor a gas engine may wish to have as much machinery as they can to save labour and time. Even in the smaller workshops such may often be wanted, and pay for itself by the saving in time. Perhaps the most useful is a circular saw which can be worked by hand or foot. Most of these machines are very feeble, and will only cut through thin stuff. These may be a convenience for some kinds of work, but one that will do more than can be done with an ordinary

142   GENERAL DIRECTIONS ON THE USE OF TOOLS.

saw in the general workshop will be the most suitable. Such a one is shown in Fig. 76, and is manufactured by

Fig. 76.—Circular Saw with Dovetailing Appliance.

the Britannia Company, by whom the wants of the cabinet-makers have been especially considered. As will be seen, it is worked with a treadle, but a handle

GENERAL DIRECTIONS ON THE USE OF TOOLS. 143

can be fitted either as auxiliary to or in lieu of this. Some idea of its capacity may be estimated from the fact that the saw can be worked at 1500 revolutions a minute, and that it will cut 10 ft. of 1 in. stuff in the same space of time. With various appliances which are made, grooving, fret-cutting, mitreing, boring holes, &c., can be done with it, so that it is a very *multum in parvo* tool.

One piece of apparatus, that shown in use in the illustration, in connexion with it is worthy of more than a passing notice. It is a simple appliance, by means of which dovetailing may be done with the utmost precision and neatness by mechanical means, requiring little or no practice. As, however, I am not treating of machinery, it would be out of place to say all that I could in favour of this and the saw, especially as full particulars can be learned from the manufacturers. To those who wish to do their own turning and fret-cutting I may commend the Nos. 4 and 10 lathes, and No. 8 fret-saw by the same manufacturers.

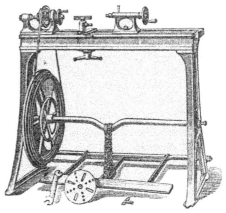

Fig. 77.—No. 4 Lathe.

The No. 4 lathe, Fig. 77, is a high-class tool, and is one of the most suitable I am acquainted with. It is well finished, and sufficiently powerful to allow of any part the cabinet-maker is likely to require being turned in it. It is made in various sizes, and is altogether a far superior lathe to that generally found in cabinet shops. For those who wish to do something more than plain wood-turning it is no use getting a light, common lathe, and it will be found more satisfactory to get a good one at once. I may say that I have examined many lathes, and have found none to surpass this one for cabinet purposes. It must be remembered that a very costly lathe, such as is required for engineering purposes, is not wanted in the cabinet shop, and this one will do any metal work that the amateur is likely to try. As the No. 4 can be had with any length of bed, it may be said that the cabinet-maker will rarely, if ever, require one more than 4 ft. 6 ins. long, while for most work one considerably shorter will do.

Although the No. 4 is by no means an extravagantly costly lathe, the No. 10 is considerably less expensive; in fact, were it not from the high standing of the manufacturers, those who are not acquainted with it might, from the price, be inclined to class it in the 'cheap and nasty' category. From prolonged use I can, however, affirm that it is not so, though of course it is not so heavy or well finished as the other, and I do not so much wish to recommend it as a first-class lathe as to inform those who wish to know of one at a very low price of its existence. It is probably the lowest-priced lathe—beyond those which are more toys than anything else—in the market, and as such is well worthy of the notice of those who do not expect to get the best at the lowest figure.

As 'toy' lathes have been alluded to, it may be said, to show the difference between them and the No. 10, that this has a 4-ft. bed and 5-in. centres. Ordinary light work can be turned in it without difficulty. When one compares its price, 90s., with other lathes, it is not easy to see how it can be sold, but that is a matter more for the manufacturers and dealers than for the user, who, however, must not be under the mistake that the lathe is the 'best.' It is undoubtedly the best at any-

GENERAL DIRECTIONS ON THE USE OF TOOLS. 145

thing like the figure, and not by any means a worthless addition to the wood-worker's outfit, though to those who do not mind the extra cost the No. 4 is recommended.

The No. 8 fret-saw is so well known that it seems almost unnecessary to do more than mention it. Though largely used by amateurs, it must not be classed in the same list as the toylike machines which are so often seen. It is a good strong tool, which with suitable saw-blades can be used to cut even 1-in. oak, though this is rather above its capacity, as stated by the manufacturers, or with ordinary fret-saw blades.

As I have said that turning and fret-cutting are separate branches from cabinet-making, I may point out the desirability of amateurs and those professionals who are a distance from turners and fret-cutters being able to do their own work in these branches. It is in any case often a convenience to do what is required, and it is by no means difficult to learn both turning and fret-cutting. For the same reasons a few carving tools may come in handy, and those who want them only for occasional work and for ordinary furniture carvings can hardly do better than get one of Lunt's sets of twenty-two of the most useful shapes and sizes. As the complete set is sold ground and handled for 11$s.$, they are as inexpensive as the common and often worthless 'amateur' sets which are frequently met with.

CHAPTER X.

JOINTS.

Squaring up—Edge Joints—Plain Gluing—Dowelling—Tonguing—Plain Dovetailing—Lap Dovetailing—Mitred Dovetailing—Bearers—Keyed Corners—Mortises and Tenons—Dowelled Frames—Halving.

THE reader may now be supposed to know sufficient to require information as to the actual work to be done. This might be conveyed by describing the construction of various articles of furniture, but it will be more to the advantage of the learner to have the principal operations involved in cabinet-making presented to him in a more systematic manner. When he can do these he ought, with the aid of a little thought, to be able to construct almost any article of furniture which he may wish.

Sawing and planing as the initial work of cabinet-making having already been treated of, nothing more need be said about them. When making anything, the importance of having the wood properly squared can hardly be over-estimated. To plane and saw at a piece of wood to get it true and square might by chance get the edges right, but to act in this way would be merely bungling, and it will be better to proceed in a rational manner.

Plane up one edge of a board straight and true; there is something to work from. To get the ends true, *i.e.*, square, is then an easy matter, whether they are merely to be squared off or the board is to be cut into shorter lengths. It is only necessary to lay the square with its block against the trued edge, and the blade across the surface of the wood, and mark across by it; this line is then sawn to. The ends afterwards ought only to require shooting to smooth and clean them, a very finely-set plane being required with end grain. The other long edge will be trued up by means of the square laid on the end, by the marking-gauge if only narrow, or, as is often done, by laying a rule across and, guided by its end, marking a line with a pencil. When getting out stuff, that is, cutting the pieces for any article, it is always advisable to cut them rather full than otherwise, to allow for cleaning off edges and ends. When marking off a board it must be remembered that the width of the saw kerf must be allowed for. It would not do to take a piece 6 ft. long and expect to be able to cut six 1 ft. lengths from it. The board might be marked into six equal parts, but when cut to the lines they would all be short, except perhaps the end pieces, if the marked line were sawn not through but by its side which is furthest from the end. The cut would then be in the wood of the next piece, and it would be correspondingly short. The same holds good when ripping also, for no one who thinks would expect to cut three pieces full 3 ins. wide from a board of 9 ins. The difference of course is not great, but quite enough to make the work defective. The trued-up edge and face of wood should be always marked. Wood, when cut and trued up, is ready for working further by joining it up with other pieces, which completed form the article.

The joints used may be divided into three classes, viz., edge-jointing, as when two pieces of board are joined together; angle-jointing, as in the making a box or drawer; and frame-jointing, for door frames, &c. In each of these classes, into which, for the sake of convenience, joints may be divided, there are several varieties, and these will now be considered. Before doing so it may be said that sometimes the choice of joint is a matter of indifference, and depends on the fancy of the worker, while in some instances a particular kind is better than others.

Edge-jointing, or, as it is very commonly called, 'jointing up,' may be taken first, and we may suppose that two boards are to be joined together to form one wide piece.

Plain Glued Joint.—In this the pieces are simply connected with glue. The edges must be shot perfectly true and square if the joint is to be both strong and neat, as it should be. It may be noted that in any joint the pieces should come close together with the thinnest possible film of glue remaining between them, so that the mark of the joint is a mere hair line, and principally observable from the different figurings of the wood.

In no case should the wood be rounded towards the end, so that the piece would be in closer contact towards the middle than elsewhere; but if the joint is a long one, the edges should be slightly hollowed out towards the middle, so that the contact would be closer at the ends than elsewhere. The least bit hollow is all that is wanted, a thin shaving taken off after the pieces are perfectly straight is sufficient. If the pieces are only short, say anything under 2 ft. 6 ins., the edges may be left straight, as in fact they must be, for either with the jointer or the trying plane it will not be easy to plane hollow on short lengths. The edges should be tried together before gluing to see that they are right; the lower piece is held in the bench-screw, the other being placed on it. Notice whether the surfaces of the boards are on the same level, and that they do not form an angle at the joint. Also draw the top one, lengthwise, a few inches backwards and forwards, and notice whether the two edges seem to work sweetly together. The feel of two perfect edges together is rather peculiar, though it can hardly be described in such a way as to render it intelligible to novices. The pieces almost seem to stick together slightly by suction.

When everything seems satisfactory they may be glued together. Of course the glue should be quite hot, and it may also be advisable to warm the edges of the wood. One piece being fastened in the bench-vice, rub the glue on and bring the edges together without loss of time to prevent the glue getting chilled. Slide the upper piece, lengthwise, slightly once or twice to exclude air and surplus glue. Then, if the wood is thick enough, apply the cramps, and mind that there is a piece of waste wood used to prevent the edges of the board being injured. Screw the cramps up tightly to squeeze nearly all the glue out and bring the edges into the closest contact. Let the cramps remain on till the glue has set and the joint become firm, which will probably be in two or three hours. Time depends greatly on circumstances when they may be removed. The joint, though the glue has set, will not be strong enough to bear rough usage for some time longer, say not till next day. If the wood is only thin, the cramps cannot be used; but when this is the case it is seldom necessary that they should be, as great strength is not required, and sufficient contact can be got by rubbing the glued edges together.

A plain, glued joint properly managed is often so strong that soft wood will split more readily than it will come asunder at the joint.

It is not uncommon to find that this joint has been used for thin sideboard and other tops, and in order to strengthen it blocks have been glued on underneath. They stiffen up a thin top if long enough, and so far are right; but the mode of application is often wrong. If a piece is glued on with its grain in opposite directions to that of the top, this cannot contract without splitting. The piece is apparently put cross-grained with the idea of binding the top so that it will not split, and is often done by the lower kind of trade cabinet-makers in London. I wish to warn readers against copying a bad example. Let the grain of the wood underneath be in the same direction as that of the top, and the risk of splitting, or in some cases of bending hollow, will be greatly reduced, if not done away with entirely.

As this is a matter of importance, illustrations, Figs. 78 and 79, are given showing the right and wrong

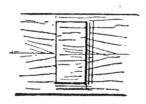
Fig. 78.—Correct.

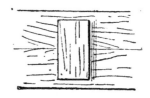
Fig. 79.—Wrong.

way of placing such blocks or linings *when they are fastened with glue.* Of course, if the pieces are merely used to strengthen a joint, they need not be so wide as shown, as a narrow piece will do equally well. The subject will be found more treated under the heading, 'Lining up.'

The *dowelled joint* is stronger than the one just described, for the reason that in addition to glue, wooden pegs, the dowels, connect the two pieces. The only difference in preparation is in connexion with these dowels, so that all that has been said about edges applies here also. Dowels can hardly be used on anything less than $\frac{3}{4}$ in. stuff, and even this is rather thin for them unless they are slighter than usual. It will easily be seen that it is of the greatest importance that the holes are bored perpendicularly and with the utmost accuracy, for if the pegs slant one way, or the corresponding holes are not exactly opposite them, the edges of the wood cannot be brought into contact. Other important and, perhaps, not such obvious points must also be observed. The dowels must fit tightly, and be thoroughly dry before using. They should also completely fill the holes bored for them, or the surface of the wood may shrink over the empty space.

The edges of the wood being ready, the exact position for boring holes must be marked. Put both pieces in the bench-screw with the edges to be joined level with each other, and remember that it is the outsides of both boards as they lie in this position which will be the one surface and the insides the other. All that is necessary, then, is to set the gauge; either the marking or cutting may be used, to mark along as nearly as may be to the centre of the edge of each board. These lines being gauged from the outside faces of both must tally. Now with the square set off lines across both pieces at intervals, say, of a foot—there is no special distance for the dowels to be apart—and so get the centres of the holes. These are thus bound to be opposite to each other if carefully bored.

A clean cutting bit should be used to bore the holes with, and none is better for the purpose than a twist bit. The depth of the holes should be uniform, and may be about $\frac{3}{4}$ in., so that each dowel-pin will be rather less than $1\frac{1}{2}$ in. There is no fixed length. To get the holes equally deep the bit gauge illustrated on page 97 may be used, or a simpler one be made out of a piece of wood with a hole through it for the bit, at any part of which it can be fastened by an ordinary screw-nail. Generally, however, no bit gauge is used, as sufficient accuracy can be got without, and as this may seem difficult of attainment to the novice, the hint that he should turn the brace an equal number of times at each hole will afford him a sufficient clue.

The holes being bored, widen their mouths slightly, so slightly that it may be considered as little more than removing the sharp edges, with the bit (countersink) for the purpose. When this has been done—and it may be explained as being partly to facilitate easy entrance of the dowels that the mouths are widened—put a little glue in the hole, and then hammer a piece of dowel wood home, cut it off at the right height with the saw, and treat all the remaining holes in the same way. Remember what was said about providing for escape of air and glue from under the dowel, and when explaining the construction of these pins. It may also be well to caution the novices against allowing the exuded glue to harden on the edges outside the dowel-pins, because it would prevent the boards coming up close. All the pegs being in, give a rub with a rasp on their edges to round the ends off a little, so that they enter the other holes easily. Glue is applied, and the boards are then to be brought into close contact as before.

The *tongued joint* is formed by one piece of wood having a projection along its centre and a corresponding groove in the other, as shown in Fig. 80. Special tools (match planes) are made for this purpose. They have not been mentioned elsewhere, for though used they are not so well adapted for cabinet-makers' as for joiners' use. For fine work they are not always convenient, as if the corners of the cutter forming the tongue get rounded it is difficult or impossible to get a close joint. In any case the joint can be prepared with plough and rabbet planes. Made as

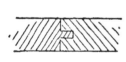

Fig. 80.—Tongued Joint.

shown in the illustration, it has no advantages over, even if it is as good as, the dowelled joint.

A much better way of forming a tongued joint is to plough a groove in each of the edges to be joined, and then insert the tonguing into each, as shown in Fig. 81. The same precautions as when boring holes for dowels must be observed for ensuring the grooves being quite opposite each other. If care be taken to set the plough properly, and to work with the fence on corresponding sides of the wood, as when gauging for

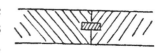

Fig. 81.—Tongued or Clamped Joint.

dowelling, it is impossible to make a mistake, as the plough iron will only cut to the depth it is set for, and the fence insures straightness.

The grain of the connecting tonguing may either run in the same direction as that of the boards or be transverse. Naturally the latter is the stronger, as it is more difficult to break across than to split with the grain. In a long joint it is impossible to have the tonguing all in one piece, so any number may be put close to each other. When this is the case, the best way to manage is to plane a piece of board to the required thickness—any kind of wood will do—then to cut lengths from the ends of the necessary width. Gluing and cramping up are managed as before.

The joint is a particularly useful one; in fact, the best for clamping up ends, that is, joining up end grain wood, either with a piece with its grain in the same direction as the main portion, as in writing-table tops with veneered banding, or as borders with the grain in the opposite direction.

Other straight joints may be and are sometimes used, but so rarely that they can hardly be considered as coming within the range of practical work.

JOINTS.

Dovetailing is the ordinary method of fastening pieces together to form corners. Of dovetail joints there are three varieties in ordinary use.

The *plain dovetail* is that in which the full thickness of both of the ends joined together is visible, and may be seen at the back of any ordinary drawer. The two portions ready to put together are shown in Fig. 82, the pins on A fitting into the sockets in B.

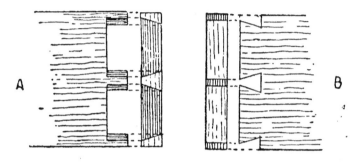

Fig. 82.—Plain Dovetails.

To make the joint, proceed as follows:—With the cutting gauge set to the exact thickness, or the merest shade more, of B, mark on both sides of the end of A. These scribings give the length of the pins. On B in a similar manner scribe the thickness of A. The beginner may then mark out the pins on the end A, and, if necessary, lines with the square on the surface to guide the saw. He will soon be able to dispense with these. Then with a fine or dovetail saw cut down to the scribed lines. The waste pieces between the pins must then be chopped out with a chisel. This should be worked from both sides of the wood, and be held so that it cuts slightly inwards from each scribe mark. An absolutely vertical cut would do, but it will be found better to work as directed, as it is easier, and all risk of a convex or rounded edge, which would prevent a close fit, is avoided. Now lay B on the bench, hold A vertically on it exactly in position, and with the marking point mark the slope of the pins, which of course should not cover the scribed lines. As before, the novice may with the square mark guiding lines, but this time across the end. Saw down to the gauged line, keeping inside the marked lines, *i.e.*, within the waste wood to be removed. To saw on the marks would, on account of the kerf, make the sockets too large. Chop out the sockets, and the parts ought to fit tightly and

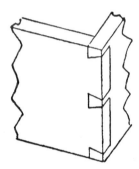

Fig. 83.—Plain Dovetail.

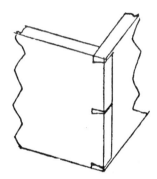

Fig. 84.—Ditto, with Badly-formed Pins.

accurately. If they do not, it is the fault of unskilful workmanship, and any defects must be remedied with the chisel. The use of this, however, should on account of the waste of time involved be avoided as much as possible. To make a really good dovetail joint expeditiously requires practice.

A method exactly the contrary of that described is often practised; the sockets are cut first, and then the pins marked from them. When this is done, A is fixed end upwards in the bench-screw, or otherwise held conveniently, and B placed on it to mark by. If the pins are cut last, remember to saw outside the marks instead of inside as before.

If several pieces are to be socketed exactly alike, they may all be sawn across at the same time by fixing them equally with the bench-screw.

In case the novice should be inclined to ask which is the better method of the two, it may be said that the answer depends entirely on the worker. Some prefer one method, some the other, and both are good. The novice may be cautioned against the very attenuated pins which are often seen. They are not recommended, those with a fairly wide thin part being better than those which taper off almost to nothing. Figs. 83 and 84 show the two kinds clearly, as well as the joint complete.

In the *lap dovetail* the ends of the piece B do not show through the other, but are lapped over. Fig. 85 gives the formation of this joint, which will be recognised as that generally used for drawer fronts. In it the thickness of B is gauged on the inside of A, and on the edge to the desired extent. The distance of this latter is also scribed on both sides of B. The rest of the work is the same as before, except that the saw cannot be used straight across the pins, which must be finished off with the chisel.

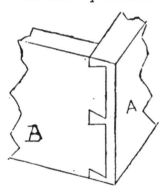

Fig. 85.—Lap Dovetail.

The *mitred dovetailed joint* externally is the same as a plain mitre. It is used when something stronger than this is wanted, but it is necessarily not so strong as either of the preceding. The general principles are the same so far as pins and sockets are concerned, but it will be seen from Fig. 86 that part of the thickness of

JOINTS.

both pieces is mitred only. The dovetails are short. Gauge the thickness of both pieces on the inside only. Then cut a rabbet of the same width and depth on each, after which the dovetails are cut and the rabbet mitred off. By a slight modification, which will at once occur to those competent to make it, both the top and bottom edges can be made to show only a mitre, or both pins and sockets can be mitred off. If the latter, the joint is much weaker, and is therefore seldom seen.

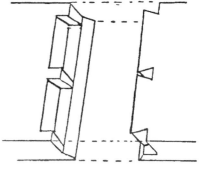

Fig. 86.—Formation of Mitre-cap Dovetails.

Any joint which is formed by a piece of wood being shaped with a dovetail is said to be dovetailed, for this form of construction is found in many instances where perhaps it might not be regarded as such by the novice. From its peculiar formation it is specially adapted to resist lateral pressure. Thus it is much used in fastening the top bearers or pieces which connect ends of carcase work afterwards covered over, as in the case of cabinets, sideboards, &c.

In this case the lap dovetail is used, the stretchers being sunk in the ends and level with their edges. When small, as in the case of pedestal writing-tables, the tops are solid, that is, covered in entirely; but if of any considerable size, these dovetailed tops, or rather top bearers, are formed of two pieces of wood of any convenient width, and generally of pine. The front bearer is faced or veneered on its front edge to match the outer wood. In order to give additional hold on the ends, these bearers are generally widened out by gluing triangular pieces to them, so that there is an open octagonal space, generally oblong, between the back and front bearers.

The construction and arrangement will be understood from Fig. 87, which shows them fitted within the ends. In addition to glue, a few nails are often used, driven through from the top into the ends to secure the bearers. If the nails are driven in a slanting direction it is impossible for the bearers to be pulled up, so that the joint is quite firm.

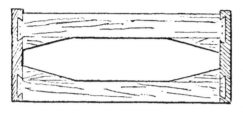

Fig. 87.—Dovetailed Top-bearers.

Bearers under drawers may also be dovetailed, but when connecting ends in this fashion, the piece of wood into which the bearer is dovetailed being both above and below, the conditions are somewhat altered, and dovetails made as any already described would not answer. Instead, the dovetail must be reversed, and be

 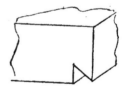

Fig. 88.—Dovetail on End.   Fig. 89.—Dovetail, One End.

formed from the edge of the wood, as shown in Fig. 88. In situations such as supposed, the dovetailing may be on both surfaces, but usually it is only formed on one, the lower, as Fig. 89. In order to prevent the dovetail showing through on the face edge of the end, and to give the bearer a square shoulder, it is usually cut back for a short distance, as in Fig. 90, and the corresponding socket stopped at a similar distance from the front edge of the end. Such a dovetailed bearer is knocked in from behind. This form is stronger, inasmuch as it binds the ends together, than the mortise and tenon or dowelled joint, one or other of which is more commonly used, and is sufficiently strong for ordinary purposes.

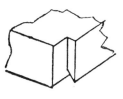

Fig. 90.—Dovetail Stopped Back from Front.

Mitred corners are often seen, but they are never used where strength is essential unless supported in some way, either by blocking inside, as in plinths, which will be found mentioned more fully elsewhere, or by keying. This is a common method for small fancy boxes, as it is strong enough for such work, and is expeditious. The edges to be joined are shot to a mitre, glued, and then further strengthened by strips of veneer, let in as follows :—Saw cuts, a fine saw being used, are made from the corners, not parallel with the top and bottom, but inclined up or down; a small piece of veneer, fitting tightly, is glued and forced in. The edge, not the end, of the veneer should be at the bottom of the cut, so that the grain is transverse. When the glue has set the veneer is trimmed off level with the surface. Such a joint appears as shown in Fig. 91, and will, no doubt, be recognised by many readers. The number of veneer keys depends on the size of the work.

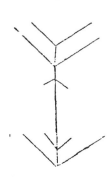

Fig. 91.—Mitred and Keyed Corner.

In framing, for doors, &c., other kinds of joints are necessary, those most commonly used being the mortise and tenon or the dowel, though much is done by simply halving the pieces.

The *mortised and tenoned* joint is much used for all

kinds of purposes, and either it or its substitute, the dowelled joint, is constantly coming into requisition. In its simplest form it is shown in Fig. 92. The tongue on A is the tenon, which fits closely within the mortise or opening in B.

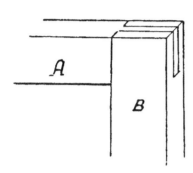

Fig. 92.—Mortise and Tenon.

To make such a joint—and, with natural modifications, all mortises and tenons are made in the same way—set the mortise gauge so that its points mark the width of the mortise and thickness of tenon, which may be about one-third of the thickness of the edge. Run it along top and bottom edge as far as necessary, and over the end of both pieces, working from the face sides, *i.e.*, keeping the block of the stop on the front. Now, with the square mark off—using a sharp edge, marking point, or chisel, not pencil—from each end a little more than the width of the corresponding piece, on edges and sides. These marks give the length of tenon and depth of mortise from the end. It then only remains to saw away the waste pieces, but unless this is done carefully the mortise and tenon will not fit tightly on account of the saw kerf. To allow for this saw on the outside of the tenon leaving the gauged marks just on its edge, and when cutting the mortise saw inside the lines. When properly managed, the parts ought to fit each other accurately, without paring with the chisel. The mortise may be chopped out with a chisel, working from both edges as in the case of dovetails. Much cutting is often saved by boring away the waste wood with a bit, and then using the chisel to square the sides and corners. This is especially the case when the mortise is a large one.

JOINTS.

In other joints of the same kind the tenon does not show at all when the joint is complete, and, needless to say, this is the neater form, though a little more troublesome. In it the tenon is shorter and smaller, being as shown by the dotted lines in Fig. 93. The mortise in this case cannot be made with a saw, but must be hollowed out with a chisel. If a wide opening, the mortise chisel should be used in preference to one of the thinner kind, which does for the sides.

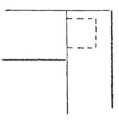

Fig. 93.—Another Form.

For wide frames two tenons may be cut, as shown in Fig. 94, and what may be considered a shorter tenon left between them, or to reduce the size of a tenon either above or below it as in Fig. 95.

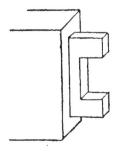 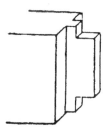 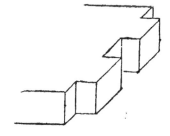

Fig. 94.—Double Tenon.   Fig. 95.—Tenon with haunch.   Fig. 96.—Tenons on End of Shelf, &c.

Such a short part is called the haunch, and is not often used in doors for furniture.

The tenon joint is often used for drawer bearers and fixed shelves. The tenons are then left the full thickness of the stuff, as seen in Fig. 96. One or more tenons may be used, according to the width of the piece, and the width of the tenons is dependent on similar considerations. Such tenons as these, of course, can seldom be allowed to appear through the wood, so care must be taken when cutting the mortises.

Tenons may be tightened up by wedging them. When they are cut through, this is simply done by driving in after the tenon is inserted, and, of course, is not practicable in such a tenon as that first described.

When the tenon is short and does not run through, it may be tightened as follows:—The end of the tenon is slightly split or cut, and one or more small wedges inserted. Their ends are left projecting, as in Fig. 97, so that on the tenon being driven home they are forced into it by the bottom of the mortise. It should be said that this method of fastening, or, as it is called, foxing, a tenon is seldom used in cabinet-making.

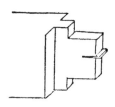

Fig. 97.—Foxed Tenon.

A *stub* or *stump tenon* is often used in connexion with thin panelling and framework for backs, as well as other subsidiary parts. In such cases, to call the hollow for its reception a mortise is rather a misnomer, though for all practical purposes it is one. Instead of being cut with a chisel it is formed with the plough, and is really part of the grooving within which the panelling is placed. Though not a strong form of joint, it is sufficiently so for its purpose.

A tenon or tongue with corresponding mortise or hollow may be used in any part of cabinet construction wherever a joint of this kind, of which the principal types have been named, may seem advantageous.

The *dowelled joint* is preferred by many, and in most cases can be used instead of the mortise and tenon. In some it is simpler and more expeditious, these probably being the reasons why occasionally writers have objected to it. Dowelling frame work is a perfectly legitimate method, and may be practised without fear. The relative merits of dowels and tenons I have no intention to discuss, for to do so would serve no good purpose. Both have their partisans, and any impartial man will admit that one form of joint is practically as good as the other. Some consider that one or other is neater, stronger, easier, quicker, or simply 'better.' All these points depend principally on what the worker has been most used to, for it is very easy to argue that dowels are better than tenons, and *vice versa*. The setting out or

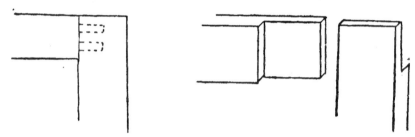

Fig. 98.—Dowel-jointed Frame.

Fig. 99.—Halved Corner Joint.

marking for dowelling follows the same rules as for dowelled straight joints—*q. v.*

In framing, two dowels should be used at each joint, one of which is represented on Fig. 98.

The *halved joint* is not infrequently used in making small cabinet doors and such-like parts, though it is hardly what one would expect to find in good work, and I do not think is ever employed in such. Still, as it answers fairly well, and is one of the least objectionable improvements (?) which are practised by cheap trade makers, it is not undeserving of notice. It is easily and quickly formed, and consists of nothing more than cutting away half the thickness from the two pieces to be joined, as seen in Fig. 99. The novice must again be cautioned about allowing for the width of the saw kerf. The joint must be glued and then secured by hand cramps till firm. Doors so made generally have the edges veneered to conceal the joints.

Halving is also useful, and may with more propriety be employed on other parts than door frames, as when two pieces cross each other in light frame work on the backs of cabinets and overmantels. The joint then takes the form shown in Fig. 100.

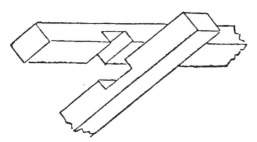

Fig. 100.—Halved Joint.

Mitred joints for frames, &c., are occasionally required, and as it is seldom that much strength is required when such is the case, the halved joint does well enough sometimes. It is precisely similar to the others, except that the face is mitred, as shown in Fig 101. Mouldings on the edges of frames, it must be noted, are always mitred, for otherwise the members could not be got to coincide.

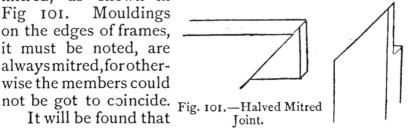

Fig. 101.—Halved Mitred Joint.

It will be found that most joints are either those described, or are such slight modifications from them, that even the novice should have very little difficulty in constructing them whenever they may be necessary. This, however, will very seldom be the case, as those mentioned are sufficient for all purposes.

In the next chapter will be described various operations which are constantly met with.

CHAPTER XI.

DECORATIVE AND MINOR STRUCTURAL DETAILS.

Lining-up—Rabbeting—Bevelled-edge Panels—Cross Grooving—Stop Chamfering—V Grooves—Beaded Edges—Stopped Beads—Flutes—Inlaid Stringing—Mouldings—Panels—Facing.

BEYOND joints there are many minor details of work which the cabinet-maker should understand, as he is constantly requiring to put them in practice, and they are as important in their way as any of the particulars which have been given. In fact, all cabinet-making consists of such details. No attempt can be made to classify such work as is now alluded to, for the different operations are mostly independent of each other.

Lining-up is the term applied to the narrow pieces which are fastened on underneath a top to give it an appearance of thickness. The lining pieces may be little more than mouldings, or they may be several inches in width. The same variety will also be found in thickness, this depending on the appearance wanted. By the use of linings a considerable saving is effected in material, for a top half inch thick may be strong enough, but its edge would look paltry; it is, therefore, made to look more massive. But perhaps some who are imbued with the false notions so prevalent about art in relation to furniture may regard such work as improper, the apparent thickness being unreal or a sham one. Well, all I can say to those to whom lining-up may be a new revelation, and who may think thus, is that they can always get a top of the same thickness throughout by paying sufficiently, and, I may add, thereby encouraging a waste of material, for to use double the quantity that is necessary without any corresponding advantage is surely this. Lining-up cannot be regarded as a sham either, for no one who knows anything about furniture would be taken in with it. Those who regard it, veneering, and such-like processes, as calculated to deceive are not generally distinguished as having any practical acquaintance with furniture construction; and, with due respect to all such would-be teachers, they may be reminded that it is not generally considered that those who are ignorant of any subject are the best guides as to what is right in connexion with it. Because *they* may have been deceived, it does not follow that others who have been better instructed are. If they have been, it has been from their own want of knowledge rather than from the perverse ingenuity of the cabinet-maker. I think my readers may safely follow the practices of skilful workers, and so I have no hesitation in recommending lining-up, veneering, and other methods with which the pseudo art-furniture critic so often finds fault.

Lining-up is satisfactory when done properly, otherwise it is likely to be a source of trouble. With the front piece there can be no difficulty, but with the ends the case may be different. These, if glued on, must be formed of pieces with their grain running in the same direction as that of the top, so that they with it show end grain. It is, however, not always so convenient to use the lining pieces thus as to have their grain running across that of the top. In this case no good cabinet-maker would dream of using glue to fasten them with, for the top being bound and unable to contract naturally would either split or, if free to do so, curve hollow.

If they are fastened with screws, and the holes in

DECORATIVE AND MINOR STRUCTURAL DETAILS. 167

the lining are made so large as to allow a little play in the top, no harm will result. The pressure of the screw-heads on the lining will keep the parts close, and yet not prevent the top contracting, as the screw-shanks will be free to move slightly backwards. Some men prefer to make slots instead of round holes in the lining for the necks of the screws, and of course from the extra provision for ensuring free play of the top they make 'assurance doubly sure.' A fairly wide, round hole is,

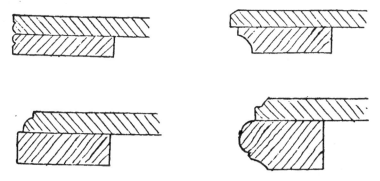

Figs. 102 to 105.—Sections of Lined-up Tops.

however, generally enough, and is more easily made than an oblong one.

The front linings may be carried right through at the ends, and the others be shouldered up against them. This does well enough when linings showing end grain are used, but not otherwise. In this case the front and end pieces are mitred, a little glue being used at the joint. The front lining may be screwed or glued only, hand-screws being used for pressure till the glue has set. It will also be as well to put one on at each of the mitres covering the joint. It is seldom necessary for a lining to be more than three or four inches wide, or to have one along the back edge. If there is a lining all round, as in the case of some tables, care should be taken to leave the

168 *DECORATIVE AND MINOR STRUCTURAL DETAILS.*

necks of the screws fastening the top to the framing sufficiently loose to allow of play, as explained above.

Perhaps the reader may discern that it is rarely that wood can be glued up with its grain running transversely across another without great risk of curvature or splitting. Let him avoid doing so, and he will be saved a good deal of trouble. It will be understood that lined edges can be treated just as if they were solid, and any mouldings be worked on their edges. A few examples are given (Figs. 102 to 105). In the case of the beaded one it will be advisable to regulate the beads so that the joint is between two of them, or at any rate sunk, and not on the face of any member. The same rule applies whatever the moulding. To make the instructions concerning end linings clear, so that there can be no possibility of a mistake, Fig. 106 is given. A shows an end lining which may be fastened in any way, while B shows the other, which *must not* be glued. The mitred and square joints will also be noted.

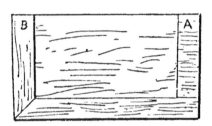

Fig. 106.—Top with Lining.

Lining-up is also used to thicken up ends of carcase work. It may be only on the front edges, or on top, bottom, and back as well. In either case the work is subject to the same rules as the other so far as grain is concerned, though it may be said that top and bottom linings should always be glued, and therefore the grain of them runs vertically. Such linings are generally flush with the edges of the wood to which they are joined. In front the joint may be concealed between beading, and by the one behind being set back a little it may save the work of cutting a rabbet in which to lay the backing.

DECORATIVE AND MINOR STRUCTURAL DETAILS. 169

Rabbeting, like much other work in cabinet-making, is simple enough when one knows how, but unless he were shown, the novice would no doubt find a difficulty in using the rabbet plane. The rabbet, or rebate, is simply a hollow or recess, square cornered, run along the edge of any piece of wood, as Fig, 107. To make it, mark its limits with the cutting-gauge, cutting as deep as convenient; then with a chisel cut away a little

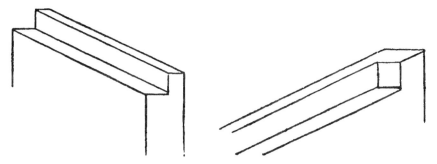

Fig. 107.—Rabbet. Fig. 108.—Stopped Rabbet.

of the wood to the line. This forms a kind of guide for the plane, with which the rest of the work is done. Hold the plane in the right hand, behind the iron, and use the left as directed for the trying-plane when working an edge. The fingers then act as a further guide to the plane, and many use no other, even when beginning to cut a rabbet. On the other hand, some plough a groove and are guided by it. The rabbet-plane may be used to cut either along the edge or along the flat surface, as may be most convenient.

Sometimes a rabbet is not required to be run through to the ends, but is stopped short, as in Fig. 108. In this case the ordinary rabbet-plane could not well be used, and the tool specially made for such cases is the 'bull-nosed' plane. In it the iron is close up to the front; it may, however, be dispensed with by working the rabbet as usual, and then filling up the end by gluing a piece of wood in. If done neatly this is quite satisfactory.

The *sunk bevelled edges*, so often seen on panels and drawer fronts, are also made with a rabbet plane, which is used much as before, the principal difference being that the work must be done from the surface, and after the required depth has been made, the plane is sloped to one side as the bevel is formed. In case any may not recognise these kinds of edges by name, they will do so by looking at the accompanying illustration, Fig. 109, where front and section are seen—both, be it noted, in rather an exaggerated form for the sake of distinctness.

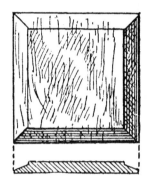

Fig. 109.—Panel with Sunk Bevelled Edges.

Grooving across the grain of carcase ends to fit shelves in them, or for other purposes, is often necessary, and, like other things, may be managed in more ways than one. In all, however, it is essential to set the places out accurately with the square from the front or trued-up edge. If the square is not long enough to reach across, use a straight edge as well. Scribe across, with the marking chisel preferably, then deepen the cut with an ordinary chisel, slightly round off the edges of the wood to be removed, as in Fig. 110. There is then a slight, straight edge by which to guide the saw. This is used with the whole of its edge flat on the wood, and a kerf of the required depth, which should have been previously gauged on the

Fig. 110.—Section of Board notched for grooving with Saw.

edges of the wood, is sawn. The waste wood between can then easily be removed with a chisel. If the groove is wide enough, and a perfectly smooth bottom is wanted, a finely set rabbet plane may be used. If too narrow for the plane, all that is required can be done with a chisel, for if the shelf is a fixed one the bottom of the groove does not require great nicety. When it is for a sliding tray the case is different, and it must be finished as well as possible. If there is a difficulty in getting the groove of equal depth throughout, the 'old woman's tooth' may be used. In some cases the preliminary cutting with the chisel is dispensed with, and the saw used directly after marking, either with or without a straight edge to guide it till it has made a passage for itself.

A special plane, called a trencher, is made for cutting cross grooves, but it is not often seen.

When the groove is only part way across, or is stopped an inch or so back from the front, the saw cannot be used altogether, though if the groove is a long one it may be partially. It must not, however, cut through the stop. The chisel must be used, and the 'old woman's tooth' will come in handy, as by the simple expedient of knocking the iron further through as the groove gets deeper a uniformly level bottom can be got at any depth required, and the cutter will work right up to the stop. Now, how about a dovetailed bearer which requires a stopped socket or groove? Well, just groove right across and knock it in from the back, or make a groove not shorter than the length of the dovetail and as thick as the bearer just behind and continuous with the socket. Put the bearer in before the ends are fastened up together, as though it were tenoned, and then push forward. To cut the socket, make a groove of the same width as the thinnest part of the dovetail, and then with the chisel cut away to the necessary slope.

172 DECORATIVE AND MINOR STRUCTURAL DETAILS.

Stop chamfering is a method of ornamenting edge by bevelling them and stopping them at certain distances. It will be recognised from the illustration, Fig. 111. The bevelling, if it were run through from end to end of the wood, would present no difficulty, but the stopping and bevelling of the stop often seem to puzzle the amateur, though really the way or ways, for there is more than one, by which they are managed are exceedingly simple. Mark off the position of the stops across the edges, and see that they are regularly disposed. The chamfering, I suppose it is scarcely necessary to say, should be done before the job is made up.

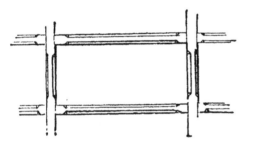
Fig. 111.—Stop Chamfered Edges.

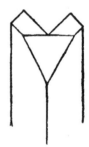
Fig. 112.—Guide for cutting Chamfer Stops

The stop will be sloped off with the chisel, and he must be a poor worker who cannot do this with sufficient accuracy. If, however, he wants a mechanical aid or guide here is one (Fig. 112), which he can easily make for himself as follows :—Rabbet out a piece of wood, or join two pieces square with each other. Any edge, or rather arris, will lie within these. Now cut one or both ends off on the bevel as shown, and use the bevel as a guide for the chisel when cutting.

As the word arris has been used, it may be well to explain its meaning, for though its use is not by any means exclusively confined to cabinet-making, it is not one of those 'familiar in our mouths as household words,'

DECORATIVE AND MINOR STRUCTURAL DETAILS. 173

It denotes the sharp edge formed by the meeting of the surface or wide part and of the edge or thickness of a board, or for that matter other things as well.

The actual chamfer, as has been stated, may be done with the scratch, which is useful enough when the amount of chamfering to be done is not sufficiently great to get a special tool, and the worker has sufficient *nous* to cut away some of the arris and so reduce the amount of 'scratching' to be done. The tool used to cut may be a chisel or rabbet plane, which will not work close up to the stop unless it be a bull-nose, when it will leave very little space to be cleaned up with the chisel. With careful management a bull-nose does as well as a special plane, which differs little from it except that it has an adjustable fence or fences on the sole. By regulating the distance between these, the bevelled edges of which fit to the edge and face of the wood being chamfered, the chamfer can be made of any size.

A tool for a similar purpose is little but an old woman's tooth with a V-shaped sole. The V of course is rectangular, and the depth or width of the chamfer is determined by the depth to which the iron projects. As any one can make this tool for himself, a front view showing shape of sole and edge of iron is given in Fig. 113. At the commencement the iron should not be down too far. This tool will cut right up to the stop.

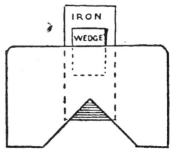

Fig. 113.—Stop Chamfering Tool.

Grooved panels, with a V-shaped grooving running diagonally, are often used in connexion with chamfered edges, and the two are taken as a plain rendering of Gothic in furniture. The cuts, Figs. 114 and 115, show a door so treated and a section of the grooving. This may either be done with a plane having a V sole or with the useful scratch, though this latter is somewhat awkward for the purpose. The grain of the wood should run with the grooving or channelling, and joints should be arranged to be in the bottom of the V. These considerations are not always observed. As the shoulder of the scratch cannot be worked along the edge of the panel to be channelled, a strip of wood must be secured to the panel to act as guide to the scratch. The work in the absence of a V plane may also be done, and perhaps more easily than by the scratch, with an ordinary rabbet plane. The channel is first partly cut with a chisel, and the rabbet plane held leaning sideways does the rest.

Fig. 114.—Door with Chamfered Edges, Frame, and V-grooved Panel.

Fig. 115.—Section of V Groove.

Beaded edges are often used with great effect, and afford an easy means of relieving monotony, for a very plain piece of furniture may have a highly decorative effect by their judicious use. They may be made with the scratch. In Figs. 116 to 119 a few suggestions are given suitable for bearers, ends, and front edges generally. The arris bead on Fig. 116 is worked in two halves, the stock or fence of the scratch being

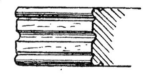

Fig. 116.—Moulded Edges.

DECORATIVE AND MINOR STRUCTURAL DETAILS. 175

worked along the edge and on the surface of the wood.

It is often desirable to stop a bead instead of running it straight through to the end of the piece. This is

Figs. 117 to 119.—Moulded Edges.

impossible with a beading plane, and the scratch, though it will stop anywhere, does not leave a clean end. This must be finished off with a chisel in order to get it clean and sharp, as in Fig. 120. The same appearance may be got by running the beads through, cutting away at the part to be stopped, and filling in the groove thus left with a piece of wood matching as nearly as possible,

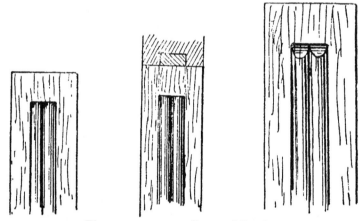

Figs. 120 to 122.—Stopped Beads.

and after the glue has set, levelling off the surface. This leaves a very clean finish, but unless the let-in wood matches well with the other is apt to look artificial.

Fig. 121 shows the details.

176 DECORATIVE AND MINOR STRUCTURAL DETAILS.

A simpler method is by bevelling off the beads with a chisel, as in Fig. 122.

Bands of beading of a different colour from the surrounding wood are sometimes seen, as satinwood beads in walnut, walnut or mahogany in ash, &c.; but unless judiciously done the effect is garish. When the beading is black, or of a darker colour than the rest of the work, it is often stained so by the polisher, but of course this cannot be done when the beading is lighter. When this is wanted, a groove must be ploughed to the required depth, which need seldom be more than $\frac{1}{8}$ in., and filled in either with strips already moulded, or plain to be worked afterwards.

Flutes or hollows are worked in the same way as beads, and are often used in decorating ends, as in Fig. 123. Small flutes and beads, when combined on an edge, are generally in workshop parlance classed as beading, though of course when merely a verbal definition is given greater accuracy is necessary. Black or darkened flutes are often seen. They are coloured by the polisher.

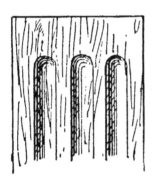

Fig. 123.—Flutes.

Inlaid stringing is much used on edges of furniture with marquetry inlays. The 'stringing' consists of bands of veneer cut to an exact width. This varies from $\frac{1}{16}$th of an inch, or even less, to a considerable size, though the larger ones may be more appropriately termed bandings, as they are generally used to finish off towards the edges of panels, tops, &c. The narrow stringing referred to will easily be recognised as forming the lines, generally of a light colour, which are often seen on rosewood and other furniture. It is kept and sold in lengths by most dealers in fancy veneers, but those who wish to do so will not have much difficulty in making their own. The veneer principally used is white or light yellow, and none is better than box. The so-called ivory stringing is seldom the real material, but either a white wood, such as sycamore, or holly, or nowadays more commonly a substance often called ivorine. It is in reality xylonite, which is sometimes known as celluloid, and is admirably adapted to the purpose. The 'grained ivory' xylonite is in appearance not to be distinguished from real ivory, and is considerably cheaper. It is more easily cut into stringings than wood veneer. Stringings are generally used in veneered surfaces. The channels may be scratched for them, or, if wide enough, the cutting-gauge may be used and the veneer between the cuts be removed with a chisel. A useful little appliance for cutting the grooves for narrow stringings is a piece of steel with a saw edge, and a wooden rim or back to form a handle.

Fig. 124.—Inlaid Bandings.

The groove is then simply sawn down to required depth. At corners stringing should be mitred ; the necessary cuts can easily be made while it is being laid. Bands of stringings are often made by having several of different colourings and widths side by side, as in Fig. 124, where, however, it is impossible to represent the colours. Artificially dyed veneers are to be had in almost any colour, though with the exception of blues and greens there is plenty of variety to be had in natural woods. 'Mosaic' bandings need only be mentioned to say that they are never used in good furniture, and seldom in any other except the commonest 'fancy' articles.

Mouldings are much used in cabinet-work. Strictly

speaking, many of the edges referred to under beading are moulded, but they are not generally understood as being included among mouldings. Mouldings are rather those portions and edges where there is a certain amount of slope varied by rounds, hollows, fillets, or flats. On edges of furniture they are usually formed by the cabinet-maker, but when they are worked in lengths to be cut from and added to the more solid construction they are often got ready made or run by machine.

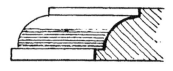

Fig. 125.—*O*volo Moulding.

If they are as large as those for wardrobe cornices the machine mouldings are much cheaper than the others, but the cabinet-maker ought to be able to make them by hand. A small moulding, such as that shown by Fig. 125, presents little difficulty even to the novice, whether it is worked with a specially formed plane used much as

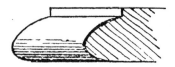

Fig. 126.—Thumb Moulding.

Fig. 127.—*O*gee Moulding.

when rabbeting, or with tools which have been mentioned in the list; it is known as the ovolo. Others much used on the edges of tops are the thumb-moulding, Fig. 126, and the ogee, Fig. 127. As shown, these are more moulded edges than mouldings.

Special planes can be had to work these and other mouldings which are commonly met with. They are known by the name of the moulding they are intended to cut—thus, that for working the ovolo is called the ovolo plane. In all of them the distinctive feature is that the sole, and consequently the cutting-iron, is shaped to the moulding. The advantage in using them is principally in point of speed, especially when a considerable quantity of mouldings have to be made exactly alike, so that to the novice they are not indispensable. By the judicious use of hollows and rounds any edge may be moulded without much difficulty, and as a special moulding-plane can only be used for cutting one size, the cabinet-maker must occasionally fall back on the simpler tools.

As an instance of the manner of using them, the formation of an ovolo may be described, as it is not only commonly met with, but will serve to show how any other edge may be worked. The rabbet-plane is first used to reduce the edge, as shown in section, Fig. 128. The outline of the moulding being marked on the end, the round is then worked with a suitable hollow, and the moulding is complete. Nothing can be simpler; and if the novice will remember to save labour by cutting away as much as he can of the waste wood with any convenient tool—in the instance named, the rabbet plane—leaving the finishing to be done with the hollows and rounds, he will have little cause to regret the absence of special planes.

Fig. 128.—Formation of Ovolo.

When working lengths of mouldings for cornice and other purposes, much the same method is pursued; indeed, the principle is identical whether the moulding be a large one composed of many members or one of the simplest character. When a large moulding is required, considerable saving in material is to be effected by making it of comparatively thin material. The amateur is generally told to cut it out of the solid. To do this entails both an increased amount of labour and a waste of wood without any corresponding advantage. Let us suppose that a moulding like that in Fig. 129 is required. If neither time nor material is an object, the worker may proceed as follows:—A length of wood, say, 2 ins. thick and 3 ins. wide is taken, the moulding marked out on the end, and as much as possible removed with the rabbet plane, either used alone or in conjunction with the plough, the final shaping being given when necessary with the hollows and rounds. Now it will be seen from Fig. 130 that this requires a good deal of work, and that nearly half of the wood,

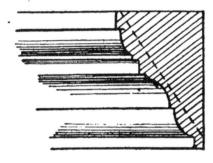

Fig. 129.—Moulding.

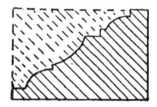

Fig. 130.—Moulding from Solid.

that portion represented by dotted lines, is waste. In such a large moulding, one for cornices, it is not necessary that the wood should be solid from back to front, so that the quicker and easier way of forming it from comparatively thin stuff is preferable. Even for large cornice mouldings it will seldom be necessary to use anything thicker than ¾-in., so that not only is there a saving of material, but the work is considerably less. Fig. 131 will sufficiently show the method, which shows the bottom of the moulding adapted for fastening on to the face of the work. When it is required to be placed on top of an edge, as in the case of a frieze—the flat part immediately under a cornice moulding—a somewhat different formation is required, and is shown on Fig. 132. Even when a solid moulding is necessary this method is

DECORATIVE AND MINOR STRUCTURAL DETAILS. 181

generally adopted in preference to that first named, the wood behind the thin facing being pine, as suggested by the dotted line in Fig. 129. Of course there may be some who do not care for this mode of construction, who think that a job should be of one wood—oak, mahogany, or whatever it is—throughout; people have been known to complain when they have discovered pine being used in furniture as parts subsidiary to finer wood. Well, all I can say is that they may use any method they please, but if they follow that practised by good cabinet-makers they will have no reason to complain. Study of economical construction is as essential to success as good workmanship, indeed the latter in-

Fig. 131.—Moulding from Board.   Fig. 132.—Alternative form.

cludes the former. Economy must not, however, be understood as being anything more than a rightful use of material. To use $\frac{1}{4}$-in. stuff instead of 1-in. does not necessarily imply economy, though it may reduce the cost in the first place. Just the same with labour. A dowelled joint takes more time than a plain glued one, but it may be more economical; and in constructing anything the efficient worker should be able to decide on the method most advantageous in any given case, having due regard both for time or labour and material.

Panels, it may be observed, should never be glued into framing, whether they are sunk in grooves or in rabbets. If glued, they will very likely split. If loose, they simply contract a little.

Panels are sometimes required to be flush on one or both sides with the framing that surrounds them. In the former the panel is rabbeted either to fit within a ploughed groove or to be fastened in a rabbet with a bead in the usual manner, as in Fig. 133. When the surfaces are to be flush on both sides the ploughed and tongued joint may be used.

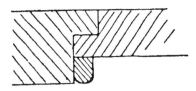

Fig. 133.—Flush Panel Rabbeted.

Facing means gluing thin wood, generally $\frac{1}{4}$-in. stuff, on to a backing of pine or other cheap material. Thus, edges of drawer-bearers are often faced up to match the rest of the outside wood of any article. Faced stuff means, broadly, that it is covered with a superior kind, of thicker substance than ordinary veneer. Facings can, therefore, be used without disadvantage where veneers would be unsuitable.

CHAPTER XII.

CONSTRUCTION OF PARTS.

Drawers—Doors—Cornices—Plinths.

SOME portions of construction and parts of furniture are of sufficient importance to warrant a chapter to themselves, as the principles and general *modus operandi* being once understood, they can easily be applied, with such modifications as may be required whenever necessary.

Drawers, and the various fittings connected with them, may receive special attention, for they are used largely, and nothing so clearly shows the skill of the workman. If he can make a good drawer, and fit it properly so that it runs firmly but easily, without either shaking or sticking, it may almost be said he can 'make anything' in furniture. The inconvenience of a badly fitting drawer, one which requires coaxing before it can be opened or shut, need not be dilated on, for we have all been made acquainted with it some time or other. In making a drawer, the front should be got out to fit the opening it is to occupy. This of course should be perfectly rectangular. The front should be made to fit as tightly as possible in the length, and in the width of the piece it should be cut slightly in excess to allow for any shrinkage. If cut bare, or an easy fit originally, by the time the drawer is made it will probably be too easy. Each drawer front, when there are several of the same size, should be marked to show the space it is intended for, and the same for all parts, that they may not get mixed up. If this precaution is taken, each drawer, even though there may be slight differences between them, will be accurately placed. From $\frac{3}{4}$-in. to 1-in. stuff will do for most drawer fronts.

Drawer sides are got out in the same way, being cut to the right length, but left a trifle full in width. Each should be fitted to the place it is to occupy, especially if the worker is only a comparative novice. A skilful worker, if several drawers are of the same size, may fit all the sides in one place and find them ultimately all right, but the novice would not be safe to do so. Drawers, it may be pointed out, rarely go right to the back, so that the ends may be cut accordingly.

Drawer backs are seldom so wide as the fronts. Their top edges are lower, and the bottom edges are above those of the other parts. In length they should be full.

For both sides and backs $\frac{3}{8}$-in. to $\frac{1}{2}$-in. stuff is generally suitable.

The lap dovetail is used in front and the plain at the back, the pins being on these pieces and the sockets in the sides.

Along the lower edges of the sides inside are glued the 'drawer bottom slips,' which are the whole length of the sides. They are grooved to hold the drawer bottoms, and though there is a special plane for doing the grooving, it is not necessary, as the work can be done equally well with the plough. For the sake of neatness, the edges of the slips are generally rounded off inside the drawer. Along the front, and in a line with those in the slips, is another groove for the front edge of the bottom to fit in. The lower edge of the drawer back is on a line with the top of the groove, so that the bottom board when pushed into its place fits just under the back. The bottom board must project at the back to allow for shrinkage, and it seems almost

CONSTRUCTION OF PARTS. 185

unnecessary to add that the grain of the wood must be across the drawer, *i.e.*, from side to side of it. Many cabinet-makers have a bad habit of fastening the bottom to the back by a nail hammered in, for of course to use glue to fasten it anywhere would do away with the advantages of the ordinary construction. A small screw nail is much better, for if the drawer bottom shrinks it does so towards the back, and may possibly pull out of the front groove. In this case the screw is easily removed, and the bottom pushed forward again.

Now with this general idea of the component parts of a drawer, the following illustrations will require very little explanation.

Fig. 134 represents the plan of a side of the drawer showing the bottom projecting behind. It will be noted

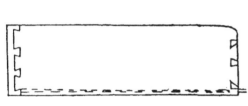

Fig. 134.—Drawer Side.

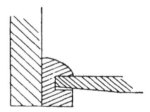

Fig. 135.—Fitting of Drawer Bottom.

that the top and bottom pins of the front are only half dovetails, and the same with the bottom one of the back, which is level with the top of the bottom board. The top one may be made either way, but is best as shown. The back corners of the sides are also rounded off to allow easy entrance of the drawer.

Fig. 135 gives a section of the drawer side, showing the position and shape of the bottom slips with the bottom fitted in. If the bottom is too thick for the groove, it should be thinned from the under side.

Fig. 136 shows the inside of one of the sides with its slip, and sections of the front, bottom, and back.

The number of dovetails may be increased or decreased according to the depth of the drawer, but it should not be less than three at each angle.

Drawers made by joiners, and even occasionally by cabinet-makers, have no bottom slips, the grooves being ploughed direct into the side.

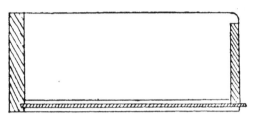

Fig. 136.—Drawer Side with Bottom, &c.

Fig. 137.—Alternative Fitting for Drawer Bottom.

Another plan sometimes adopted, but rather an amateurish one, is to glue one slip on below the bottom and another above it, as in Fig. 137. This practically forms a groove for the bottom. Very small drawers, as those in portable writing-desks, often have the bottom glued in, for little fancy things are not always constructed exactly the same as larger articles.

When the drawers are long, the bottom, in order to give it extra stability, is made in two or rather three pieces, the centre one being a munting, which may be explained as being a wide bottom slip with a groove on each side into which the bottoms proper slide. This centre piece, though necessarily thicker than these, is thinner than the depth of the slips. In front it should be tenoned (stub tenon sunk in the groove), and screwed behind to the back. It is sufficiently shown in Figs. 138 and 139, the former being in section with bottom adjusted, the latter in elevation with section of bottom and back.

If there were only a single drawer in a perfectly

CONSTRUCTION OF PARTS. 187

plain box, it would of course run against the bottom, top, and sides; but with several drawers one above another the case is different. There are rails between the drawer fronts, and others extending from back to

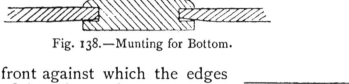

Fig. 138.—Munting for Bottom.

front against which the edges of the sides work.

The bearers in front are fastened into the carcase ends by tenon, dowel, or dovetail, and may usually be anything from 3 ins. to 6 ins. wide.

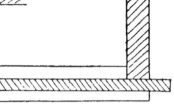

Fig. 139.—Ditto, showing Back.

The front edges are faced, as whatever wood the job is made of pine is generally used for bearers. On their upper surfaces are nailed the drawer stops. These are merely pieces of wood, one or two for each drawer according to size, thin enough for the drawer bottoms to pass over, and so placed that the fronts stop against them. The runners for the drawer sides are tenoned into the bearers, and may be of any suitable width, the thickness of bearer and runner being exactly alike. In fixing

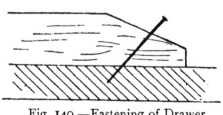

Fig. 140.—Fastening of Drawer Runner.

them, beyond the obvious necessity of having them square across, the principal point to be observed is not to glue them on to the ends. If glued, these would have no opportunity to contract. The tenons are glued into the mortises, and a touch of glue may be used just by the bearer to fasten them to the ends. At the back they are merely held by a nail, either a screw or other kind, driven in on the slant, and left slightly projecting, as in Fig. 140. If the end shrinks it can then do so without risk of splitting.

Extra strength when necessary may be got by grooving the ends and sinking the edges of the runners in them, but is seldom necessary.

The inner edges of both bearers and runners are often grooved for a thin board to be inserted, as it easily may be, from the back. This 'dust-board,' though seldom necessary in chests of drawers, should not be omitted when the lower part of the carcase is a cupboard. If it is, by the simple expedient of taking the drawer out a dishonest person can obtain access to the cupboard even if its door is locked.

When the drawer sides fit closely against the ends of the carcase, this is made a shade wider at the back than at the front, *i.e.*, a trifle out of the square. So little, however, that the divergence represented by the thickness of a piece of veneer is more than sufficient. A 'shaving' more exactly implies what is wanted, as the only object of this construction is to diminish the friction of the drawer sides and ends. Some may think that to make the drawer sides slightly thinner towards the back would do equally well, or even that the drawer might be slightly narrower at the back than at the front. Either of these might do with inexact workmanship, but not otherwise, for reasons which will soon be discovered if tried.

When the carcase ends are thickened up in front as well as between drawers working on the same runner, as in the case of the two short ones which are usual in an ordinary chest of drawers, an addition must be made to the fittings to prevent the drawers working sideways. These guides take the form of rectangular strips of wood fastened on to the tops of the runners, and may be of any convenient width and thickness.

The arrangements of the various parts will be recognised by the illustrations. Fig. 141 shows the plan, Fig. 142 the section; the lettering in both being the same. Drawer fronts may be either plain, have sunk bevelled

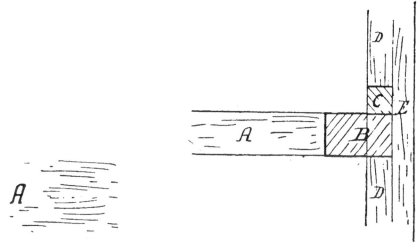

Figs 141 and 142.—A, Bearer. B, Runner. C, Guide on top of B. D, Lining of end E.

edges, be beaded across or be finished in other ways. Perhaps none is more effective than that by means of

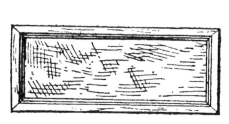

Fig. 143.—Drawer Front and Moulding.

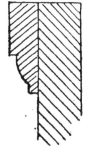

Fig. 144.

small mouldings planted on round the edges, as shown in Fig. 143, of which the section is given in Fig. 144.

Drawers should not project beyond the bearers and

ends, but should rather be set a trifle back, say, one-sixteenth of an inch when closed. Those with stuck-on mouldings should be set back still further, as it is rarely that the mouldings should project.

Doors.—The principal parts of these are the frames and panels. The upright portions of the frames are the stiles, the cross-pieces being the rails. These are fitted within the former, that is to say, the stiles run the entire height of the door, and the rails lie within them. The reverse is sometimes seen in old but never in ordinary modern work.

The panels may be sunk in grooves, and must then be let in before the framing is made up. The usual practice is, however, to sink them in rabbets from behind, and then fasten them in with beads, as shown in section, Fig. 145. The

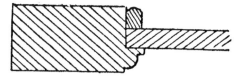

Fig. 145.—Panel Rabbeted in Frame.

beads should not be glued in but be neatly fastened with small wire nails, or better still, with screws, which should be round-headed brass when appearance is an object.

The frames of doors are often not rabbeted in the ordinary way, but after they are made up small mouldings are glued on the edges near the face sides, as shown in Fig. 146. These mouldings, as will be seen, form the rabbet, and are often highly ornamental in effect. In other cases the frame is rabbeted, and the mouldings worked on the solid. So far as appearance goes, and even utility, there is little to choose between the two.

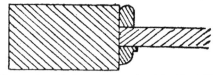

Fig. 146.—Rabbet formed with Moulding.

CONSTRUCTION OF PARTS.

When the moulding is worked on the frame, *i.e.*, not stuck on, it must of course be mitred and cut away to allow of the joint being made, as shown in Fig. 147. The loose mouldings are simply mitred and glued in.

If the moulding is of any considerable size, as is

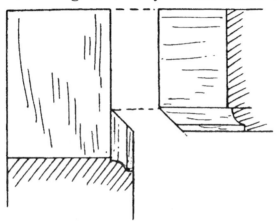

Fig. 147.—Mitred Mouldings Worked on Frame.

sometimes the case with large doors and other frames, it is advisable to sink it in a rabbet, especially when the panel is looking-glass, as greater strength is gained. This method is illustrated in Fig. 148.

All mouldings on the edges of frames should be thoroughly dry, for if not, they will shrink and become open at the mitres, where, without a great deal of trouble, it is impossible to rectify joints. It may be a fact worth mentioning to some as a curious one to them that an opened mitre is not caused by the moulding shrinking in length but in width. Wood, it will be remembered, does not contract in the former direction.

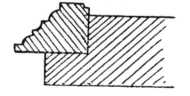

Fig. 148.—Moulding in Rabbet.

When arranging the rabbets of a frame see that they are of equal depth from the back on all the four edges in order that the panel may fit equally to them. It may also be well to see that they are not winding, as a very trifling irregularity at the joints may make them defective in this respect as well as out of the square.

The stiles should be left long, so that they project at each end a little beyond the rails till these have been fitted in. This precaution is especially necessary when the tenon joint shown on p. 161 (Fig. 93) is being used, as were the pieces cut exact there would be more risk of splitting when cramping the tenon home.

As the rabbet may cause some embarrassment to the novice when framing up, he may refer to Fig. 149, showing the appearance of a top end of a stile with the rail. He will then have no difficulty in knowing how to proceed, as part of the rail is cut for the rabbet.

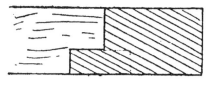

Fig. 149.—End of stile showing rabbet.

The novice must be cautioned against adopting a method sometimes practised, especially by amateurs, for it seems such an easy way of getting the appearance of a framed door. It is simply by gluing pieces of wood to represent the rails and stiles on a panel the full size of the door. One so made is nearly certain to cast, as either the rails or stiles must be across the grain of the panel. It may be taken as a general rule that simple looking makeshifts are not generally of any practical use, for otherwise they would be employed by skilled cabinet-makers, who have every inducement to work in the simplest effective manner.

Cornices and *plinths* enter so largely into furniture construction that they may receive attention here, though perhaps it may seem rather incongruous to class them together, forming, as they do, the extremities, top and bottom, of the pieces of furniture of which they form parts. Their construction is, however, very much the same, so that for all practical purposes they may, to some extent, be taken together. They are either fixed or loose, generally being the latter if of any considerable size, as in the case of large wardrobes. In these and similar pieces of furniture formed by joining more than one carcase together, they serve to some extent to bind them, and are thus useful as well as more convenient

Fig. 149a.—Plinth or Cornice showing Fastenings.

than when fixed. Indeed, in many articles they could not be made in this way. To a certain extent both cornices and plinths are alike, being merely frames with the width of the wood showing, or shallow boxes without top or bottom. For articles to be placed against a wall, only the front and two ends are presentable, the back rail being merely a piece of pine fastened in between the two ends to steady them. It may be fixed by blocks glued in, but a better way is to use a dovetailed joint as described for drawer bearers. To allow of the sockets being cut the piece is fixed a few inches from the back. Fig. 149a shows the framework of such a cornice or plinth. At one end it will be seen that the back is represented fixed with blocks, and at the other as recommended. Such plain frames require adapting to the piece of furniture they are to fit, and, though impossible to go into all details here, a few hints will prepare the way for adaptation to special cases. The front corners should be neatly joined, and generally should be mitred. The mitre dovetail may be used, but an equally suitable and quicker method is to simply mitre the ends and then block them in behind. This may not seem a very strong way, and occasionally it is not strong enough.

In such cases the most satisfactory way is to dovetail two short pieces of pine together, the width of the wood being slightly less than that of the cornice or plinth. These are then fastened in the corners with glue and screws, and, if necessary, further strengthened with a block. This, however, is seldom necessary, as 1-inch pine, or even less, is, when properly joined, ample. Corners so fastened are represented in Fig. 149a.

In the case of cornices a moulding generally finishes off the top. When made as represented in Fig. 131, it may be glued to the face of the frieze, and fastened on more securely with a few screws from behind. Short blocks may also be glued on to give extra stability.

With a moulding finished as in Fig. 132, it is simply glued on to the top of the frieze, and fastened with glue and screws, as in Fig. 150, where the pocket or hollow cut for the reception of the screw is indicated. At the corners the mouldings are mitred and blocked.

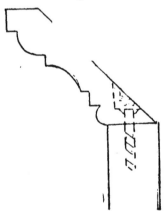

Fig. 150.—Cornice Moulding.

The top edges of plinths generally require a moulding of some kind to be worked on them, and are seldom

CONSTRUCTION OF PARTS.

thick enough to project sufficiently when the carcase part rests on them. What is done then is merely to line them up so that the section is as shown in Fig. 151. These pieces are mitred at the corners, and seldom require any other fastening beyond glued blocks. It is sometimes sufficient, especially when the plinth is a fixed one, just to fasten it direct to the carcase by glued blocks or screws, and to omit the pieces which form the occasion for Fig. 151.

Fig. 151. Lining of Plinth.

When the plinth or cornice is movable it is generally kept in position by blocks or pieces of wood fastened to the top or bottom of the carcase. In either case these are arranged so as to fit exactly into the corners, so that plinth and cornice are not free to slide about and be displaced without lifting.

An expeditious and strong method of forming a plinth or cornice, or, perhaps, I should rather say giving the appearance of these parts, is to let the ends of the carcase run through, and on these fix the cornice moulding or ends of plinth. The front can easily be added, if necessary thickening it up. In this construction it will be noticed that the work really stands on the end pieces, so that on the score of strength there is nothing to be objected to, while the appearance is exactly as before.

CHAPTER XIII.

GLASS IN FURNITURE.

Sheet Glass—Plate Glass—Purchasing—Flaws—Bevelling—Silvering—Measuring—Fixing.

GLASS, either transparent or silvered, *i.e.*, looking-glass, plays an important part in modern furniture, and though, on the whole, it may be regarded and treated as wooden panels, there are special means used to fix it which the cabinet-maker should be informed about.

Before describing the methods of fixing, it will be well for the benefit of the novice to say something about the varieties used, and give a few hints about measurements.

When transparent, glass may be either ordinary sheet or plate, and for small panels the former does fairly well, though when price is a secondary consideration to quality, the latter is preferred. Sheet glass is much thinner than plate, and of inferior quality generally. The thicknesses principally used are known as 15 ozs. and 21 ozs., the former being the thinner of the two. When sheet is used it should be carefully selected in order to get it as free as possible from flaws, for glass that would be quite suitable for windows might not be good enough for furniture.

Plate glass is thicker and better adapted to the cabinet-maker, the only thing that can be said against it in favour of sheet being its weight in large doors, though this is not a serious objection if these are properly made and hinged. It is the only kind that is or should be used for mirrors, for which the best quality only ought to be selected. In common furniture little regard is paid to this, and even silvered sheet is sometimes seen.

Glass is sold by superficial measurement, and can be had in larger sizes than the cabinet-maker is ever likely to require. There is no occasion to stock it, as the best and usual plan is to order it as wanted, ready cut to special sizes. In fact, all the cabinet-maker has to do is to fit it into its frames. The glass merchant, silverer, or beveller does all the rest, and on him the purchaser must chiefly depend for quality. Some facts about glass, however, will not be uninteresting to the user, for even if a knowledge of them is not essential, it may save the beginner both annoyance and expense. With the manufacture of glass we have nothing to do, though on account of objections which have been raised against its use in large sheets in furniture by some writers, it may be well to say that it could till comparatively recent years only be made in small pieces. This is sufficient reason for it having been used sparingly by those older furniture designers who are sometimes held up as models of all that is correct. By the way, Chippendale & Co. did not use iron and brass bedsteads, *ergo*, by a similar argument, these are not suitable for modern bedrooms. Just the same with glass. Of course, an excessive amount of glitter is hardly an evidence of good taste, but this is a subject which can hardly be discussed here. Suffice it to suggest that what is in good taste to one is barbaric splendour to another. All ladies are supposed, or suppose themselves, which is quite another matter, to possess good taste, and we find them differing entirely in their views as to 'what's what' in furniture. If it were not heresy to do so, I would hint that they mostly act as they would with dress, viz., follow the fashion. With such examples as these before him, the cabinet-maker cannot do better than use his own discretion about glass, only he need not be led away by the notion that because old furniture has very little in it his must not have.

In most of the largest towns there are dealers who supply glass for cabinet-makers, *i.e.*, it is specially selected by them for the purpose of being used in furniture. Where there are not such dealers it is no use going to the ordinary glazier or builder, especially for silvered glass, though the quality of the metal he supplies may be equal, *i.e.*, the sheets themselves are of the same make. If ordinary window glass has a few flaws it does not matter, but in a silvered plate every scratch or air bubble becomes conspicuous. On its freedom from such defects the quality of plate glass principally depends for silvering. Some importance may also be attached to colour, and this will be seen to vary if the plates are critically examined. An absolutely colourless glass is hardly obtainable in the form of plate, and for all ordinary purposes any good plate, whether of English, Belgian, or French, is practically without colour. Each kind has its own peculiarities, French being generally the whitest, while English has a greenish hue. The former, however, is more apt to change than the latter, which alters very little, and is therefore preferred by many. In any case the difference is so trifling between the different kinds that few, unless experts, would observe it.

It may be interesting to note that the principal dealers arrange their plate glass into three classes— ordinary glazing, best glazing, and silvering. The classification, however, is more or less a chance one, as there is no fixed standard to determine by. The best are put in the latter class, those not so good go in the second, while the first one includes the rest. Everything, therefore, depends on the selector, so that it is of considerable importance to the user to deal with a respectable firm, or he may find a very inferior plate classed as silvering quality. It will be found that the average of the best glazing supplied by some is quite equal to the silvering quality of others. On looking at any dealers' lists it will be seen that there is a good deal of difference between the prices of the three classes. Generally, however, the cabinet-maker buys his glass ready silvered, and it is then assumed to be only the best kind. This, however, is not invariably the case, and the purchaser should carefully examine each plate all over for flaws. These are principally scratches caused by careless handling, air bubbles, or seeds in the glass, and water stains or marks either on the silvered surface of the glass or on the surface of the silvering. If, as is sometimes the case, a defective plate is issued, in spite of ordinary precaution by the silverer or dealer, no reputable house will refuse to admit the complaint if a reasonable one. The user, however, must not expect to get a really perfect plate or one that some fault could not be found with. The difficulty of getting one without any flaw at all increases with the size of the glass, so that the novice must not fancy he has been unfairly treated if he finds some blemishes. Possibly, if he looks at a bevelled edge silvered plate closely, especially if the glass is laid flat with a bright top light shining down on it, he may be surprised to find it almost covered with very minute scratches. These are not what are referred to, as it is practically impossible to get a bevelled plate which does not appear so under the circumstances. The marks will be invisible when the plate is in its proper position. Scratches to which objection can be taken are very different, and are caused by careless treatment. The same may be said of the water stains. These are rather difficult to describe. They very much resemble the appearance of a drop of water not quite clean having dried on the glass, and cannot be mistaken. Seeds are defects in the plate, and cannot be got rid of.

Scratches, if not deep, may sometimes be ground or 'blocked out,' but this must be done by one accustomed to the work. Defects in the silvering can only be got rid of by resilvering, which, of course, is not done by the cabinet-maker. Any plate with more than a reasonable number of seeds should be rejected. In provincial towns the purchaser must often take what he can get, but in London there are many dealers actually doing both silvering and bevelling on their own premises. Prices it will be found, on comparing lists, vary enormously, even among the London houses, and it does not always follow that the dearest means the best, though no one with an atom of experience would expect to get the finest qualities from those who habitually quote bottom figures. As a rule, glass can be bought more advantageously in London than in the provinces, where in many instances the prices are decidedly high. As every now and again amateurs announce they have made what to them is a discovery, viz., that some wholesale dealer, trade beveller or silverer, or whatever he may be called, will execute a small retail order: of course he will, as all bevelling is done for 'the trade,' and all is fish that comes to his net; but it does not follow that the occasional or retail buyer is charged trade prices—probably not. He need not, therefore, go to much trouble to find a trade silverer or beveller if he can more easily get his wants supplied otherwise. Possibly some of the larger firms might decline a retail order, but I do not know of their existence. If the retail buyer meets with any difficulty in obtaining glass occasionally, it may be useful for him to know that he can get it through any good cabinet-maker. Even allowing for the profit to the latter, it will probably cost him no more than if he bought it direct, and other things being equal, there are general advantages in getting what one wants in his own locality.

Glass and silvering are reckoned by superficial measurement. If the buyer has a price list he can calculate the cost himself. Fractions of an inch are generally reckoned as full inches, so that they are of importance; and an appreciable saving may be effected by watching them carefully, both when setting out work and when measuring for glass. Thus, a plate $36\frac{1}{8}$ ins. × $36\frac{1}{8}$ ins. would be reckoned as a 37 in. square, totalling up to 9 ft. 4 ins. super, while if a 36 × 36 could be managed with, it would contain only 9 ft. The saving here in actual measurement might not be much, but then the fact that the rate per foot increases considerably with the size of the plate must be taken into account. For the sake of an inch or two in the superficial contents the price may be 2*d.* or 3*d.* per foot higher than if it had come just under the next lower footage rate. On a large plate this is an item of consequence which the cabinet-maker will do well to make a note of. Thickness of glass is not taken into account unless it is exceptional. Very thick glass, however, is never required in furniture, though occasionally the thin patent plate is necessary. It is expensive, and only used in small pieces where the thickness of the ordinary kind would render it useless.

Bevelling is charged for according to the width of the bevel at per foot run, that is, the measurement all round, so that on a plate 1 ft. 6 ins. × 1 ft. there would be 5 ft. of bevelling. The width of bevel varies by $\frac{1}{8}$ths of an inch. It is usual to charge extra for bevelling plates above, say, 10 ft. (super), and also for those with fancy corners or shapes, such as shown by Figs. 152 to 154, as with these there is considerably more risk of breakage than with straight, square-cornered plates. For ordinary purposes, bevels $\frac{7}{8}$ in. to $1\frac{1}{4}$ ins. are useful widths.

When getting quotations it is very usual to get them

for the silvered bevelled plate complete, without specifying rates, but the inquirer must state exact size of glass and width of bevel required.

The silvering by which the reflecting property is given to glass is of two kinds—the old-fashioned mercurial process and the 'patent' process, in which a thin film of silver is precipitated on to the glass by chemical action. In the former, which is now seldom used, an amalgam of quicksilver and tinfoil is caused to adhere mechanically. It is, compared with the newer process, costly and tedious. For a considerable time there was

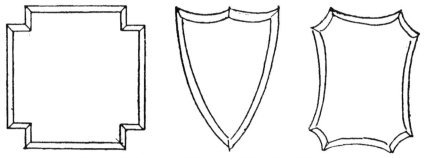

Figs. 152 to 154.—Shapes of Bevelled Glasses.

a good deal of prejudice against the new process, but this has almost entirely died away, for in the result it is found equal to the other. Although familiarly spoken of as the new or patent process, it has been in common use for many years, and anybody may practise it. In the mercurial process the metal is visible behind the glass, but with the silver one it is protected by a kind of paint which allows it to be much more freely handled than the other. The silvering by means of it can now be done in a few hours instead of days, as used to be the case. It is, however, well to order the glass, especially if it is to be bevelled, as soon as its size can be determined from the job itself—not from the drawing, as there may be trifling differences in this—so that it may be ready when wanted. It must be remembered that the work of selecting, cutting, bevelling, &c., has all to be done after the glass-dealer receives the order, and this generally takes from a week to a fortnight to execute. As a matter of favour, plates may be got in two or three days sometimes.

As I frequently receive letters from amateurs asking how to do silvering, I may take this opportunity of telling my readers that it is not suitable work for them to take up by either process. To manage properly on any but the smallest bits of glass, special appliances are wanted, and a fair amount of experience, which can only be got practically. For those untrained to the work to attempt silvering is only to court failure. Even if they could manage to do it satisfactorily there would be no pecuniary saving, as the silvering rates are almost ridiculously low.

Convex mirrors need only be mentioned because they are sometimes supposed to be old and not made now. This is a mistake, as they can be got from almost any glass house. They are not much used.

When measuring for glass, the size of the opening, or what is known as 'sight' size, should be carefully taken, as, of course, nothing smaller than this will do. If the plate has plain edges, *i.e.*, unbevelled, it may be got large enough to fit closely within the rabbet, without leaving any open space round the edges. However, as this might run it above a full inch and no advantage is gained by having more than sufficient within the rabbet, measurement should should also be taken of the full or rabbet opening behind. For all practical purposes, a quarter of an inch hidden is all that is required, though, beyond measurement, there is no reason why it may not be considerably more. To prevent unpleasant reflection from the rough edges, these should always be blackened. A mixture of thin glue and gas black is as good as anything. The front of the rabbet, against which the face of the glass lies, should also be blackened with the same or a similar mixture. This blackening only refers to plain-edged silvered plates, for if they are bevelled it is not necessary always, though there can be no objection to its being done, and some make a practice of treating all silvered plates alike in this respect.

Measuring for bevelled plates requires more accuracy, for, as the cost of the bevelling depends on the width, it is folly to have more than sufficient hidden. I may just refer to the erroneous notion sometimes held that the more of the glass that is within the rabbet, the more secure the plate will be. If the glass lay flat on the wood there might be something in this, but on account of the bevel it can only come in contact with the arris of the rabbet edge. What then is enough? Well, the ordinary custom is to make a bevelled plate $\frac{3}{8}$ in. longer and wider than sight size. This practically allows rather more than $\frac{1}{8}$ in. behind the rabbet, the remaining $\frac{1}{16}$ along each edge being to provide against contingencies of rough edges. When ordering bevelled plates it is always advisable to state whether sight or plate size is given in order to prevent mistakes. If ordered sight they will be supplied as stated, $\frac{3}{8}$ in. larger than the measurements given. For any except rectangular straight-edged plates measurements alone are not sufficient; an exact template or pattern drawn on a piece of paper or board should be supplied. It can easily be made by putting the wood or paper in the rabbet and ruling round the edge of the opening. This line should be described as 'sight,' otherwise the cutter is very apt to cut the plate exactly to it.

Transparent plates are more convenient when they fit exactly to the rabbet opening, but still there is no use wasting the bevel. If the rabbet is too wide it is better to fill it up with wood neatly fastened in.

GLASS IN FURNITURE.

The usual way of fastening silvered plates in is by means of glued blocks. With plain-edged plates no special precautions are necessary. The frame or part being glazed is laid face downwards on the bench top— see that all tools, nails, &c., are removed from within the opening—and the glass placed. If it fits close, a few blocks glued above it will prevent it falling backwards, and also prevent the wood backing, which is put on afterwards, from coming in contact with it. When gluing blocks in, be careful that the glue does not get on the back of the plate. If any drops fall on it, they should be wiped up without delay, for if left to harden, they are very apt to pull the silvering beneath them away from the glass. The blocks themselves may be from two to three inches long, and should be placed at intervals, say, of six to twelve inches, according to the size of the plate. Fig. 155 shows a tightly fitting plate with block, which if the backing is sunk within

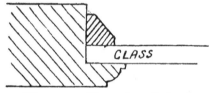

Fig. 155.—Glass close fitting.

the opening must be below the level of the back surface. If, as sometimes happens, it is not convenient to have the rabbet deeper than just sufficient to hold the plate, or the back board comes directly in contact with it, it is advisable to have a few sheets of soft paper or a sheet or two of flannel between them. With very small plates this is not necessary, though with large ones it is a safeguard against injury. With plates silvered with mercury, the flannel should not be omitted if there is the slightest chance of the blind-frame—the backboard—coming in contact with them.

It is, however, comparatively rarely that the glass entirely fills the rabbet, and then wedge-shaped blocks, as in Fig. 156, are most conveniently used. These, when firmly glued in at intervals all round, prevent the plate moving in any direction. As the blocks are not always fitted with the greatest nicety, for which there is no occasion, a wire nail is often driven through them slantingly into the frame. When this is done, care must be taken that the nail does not chip the edge of the glass.

Fig. 156.—Glass not close fitting.

With bevelled plates more care must be taken to adjust them closely, so that an equal width of bevel shows all round in front and that the mitres are exactly in their corners, for nothing looks worse than to see these all awry, and a wider bevel along one edge than elsewhere. In order to fit them accurately, it is advisable to have the frame so supported that the fitter can look underneath and see how the glass lies. If it is not quite in place, do not be tempted to move it with a chisel or screwdriver or any tool in fact, used like a lever. If this is done, the edge of the glass, especially if it is a large and consequently heavy one, is very apt to be chipped. With small plates there is little or no risk; but whatever the size, it is better to adjust by slightly

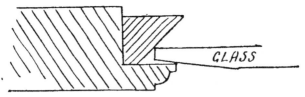

Fig. 157.—Bevelled Glass blocked.

hitting the edge of the frame with the hand. This will cause the slight movements necessary to the plate. When it is satisfactorily placed, put the wood blocks in their places without glue. This will prevent the

GLASS IN FURNITURE.

plate being moved as each one is glued. The wedge-shaped blocks are almost invariably used with bevelled mirrors, and if there is a wide rabbet the lower edges must be cut, as shown in Fig. 157.

Transparent plates, whether plain or bevelled edged, of course cannot be fitted in with blocks. Putty may be and sometimes is used, but it is by no means a nice way of doing the work, even though it is coloured to match the wood. An altogether superior way is to fasten the glass in as if it were a wooden panel, viz., with beads, as shown in Figs. 145 and 146 (p. 190). As has been said, the glass should fit closely, but if it is a trifle loose it may be secured by thin strips, which will be hidden by the beads. If, however, the space is too wide to allow of this, as may easily be the case with bevels, the best way is to neatly fill the rabbet by gluing a stop along each side and then fastening the bead on this, as in Fig. 158.

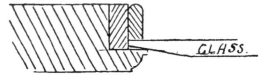

Fig. 158.—Transparent Bevelled Glass fitted.

When fitting a plate, it should always be so arranged that any flaws may be as inconspicuous as possible. By a little judicious management of this sort they may often be quite undistinguishable when the glass is fixed up. It is, therefore, well to see what can be done with any plate before condemning it. Thus we may suppose an overmantel plate with some defect near one end. If this were placed at the bottom, the faults could hardly escape observation, but at the top they may be out of sight.

Glass is also sometimes used for shelves. When this is the case, any edges which may show should be ground and polished to give them a good finish.

With these directions the novice ought to be able to fit and fix any glass in any piece of furniture in a workmanlike manner. This chapter may be concluded with the hint that any frame intended for glass must not be in winding. Glass is flexible to a small extent and might be forced to fit the winding, but in all probability the result at some time or other would be a crack across it. Also see that the rabbet is clean and free from hardened drops of glue, &c., and on the same level all round before the glass is laid.

CHAPTER XIV.

DRAWING AND DESIGNING.

Considerations for Guidance in setting out Work—Miniature Designs—Working Drawings.

WHEN working from a small sketch or when attempting to design furniture, the amateur is often at a loss to adjust the main proportions and sizes of the different parts, as well as to determine the thickness of stuff for them. To say that there are no general rules which are adopted, often, no doubt, unconsciously or in accordance with custom, by designers of furniture would be hardly correct; but they are so vague that I do not think any attempt has been made to formulate them. They are more general principles than rules, and the chief of these is unquestionably fashion or custom, though underlying this will often be found considerations of convenience.

These two principles, then, may be looked on as being the keynotes in designing, or, perhaps I should say, setting out furniture, for designing rather indicates devising or arranging details, while reference is being made now more to the size of the whole thing or of parts. If the novice asks which consideration, that of convenience or of fashion, should be the first, I say, unhesitatingly, the former, for the latter is more or less evanescent. Important, no doubt, if we wish to give the furniture a familiar, homelike appearance, but in no way affecting its utility.

Thus we find an ordinary dining-room chair is always

about one height, whoever made it. That height has been established by custom as being generally convenient. To set out a chair, therefore, as a *dining* chair, two or three inches higher or lower than usual, would, however beautiful the design otherwise, be unsatisfactory. The thing would not be adapted to its ostensible purpose. Of course, I have nothing to do with those peculiar individuals—perky Lilliputians, or long-legged giants—who, without having given a moment's serious thought to the subject, think all furniture would be much better if adapted to their special sizes. Easy chairs, on the same lines, are naturally lower than ordinary chairs.

In the same way, we find that the dining-table is made by all to the same height, or very closely. It is recognised as being best, or, what is much the same thing, we have grown accustomed to it, so that any other would be awkward and strange.

Writing-tables, in the same way, are made to a fairly uniform height, and the same with tables which are sat at in ordinary chairs. On the other hand, 'five o'clock tea-tables,' whether Sutherlands or distinctly 'fancy,' ladies' work-tables, &c., used in conjunction with low drawing-room chairs, are lower and of very variable height. They are the occasional things. Ordinary table height may be considered as 2ft. 6 ins., sometimes a little more, sometimes a little less. Three inches one way or the other would make the table either high or low; specially useful, perhaps, to a tall or a short person, but to most of us inconvenient.

A few more articles of furniture may be named by way of example, and it will please be noted that ordinary English furniture is spoken of, not extraordinary specimens.

The bookcase is one of them—the kind with glass doors in the upper cupboard, the lower part projecting a little in front, and having wood-panelled doors. Now, the primary intention of a bookcase is to hold books. The shelves, therefore, must be wide enough for the ordinary sizes, and as comparatively few books are over 9 inches wide, a little more than that is generally sufficient width for the shelves. As the heights of books vary, there is a reason for movable shelves which can be fixed at any distance apart. The total height of the case varies, but not to any very great extent. Appearance must be considered in some degree, and extra height gives dignity, but roughly the height should not be so great that a man cannot either standing on the floor, or with the aid of a 'library step-chair,' reach the top row of books. In libraries where the wall is lined from floor to ceiling, other circumstances govern the arrangements, and these may be noted as the exceptions.

On similar lines, the music cabinet should be wide and deep enough for its drawers and shelves to hold music of the ordinary sizes. Usually, for the sake of appearance, it is made somewhat larger, for one just large enough to hold music sheets looks paltry. Wardrobes also may be, and often are, made greater in height, for appearance' sake, than there is any absolute necessity for. A very high wardrobe, be it noted, is not necessarily the most convenient for the user, for some of the hooks, shelves, or trays may be out of easy reach.

Perhaps the most unsatisfactory all-round piece of furniture is the hall-stand, with accommodation for hats, coats and umbrellas, looking-glass and drawer or box for brushes, gloves, &c. There is in it an attempt to combine too much, and a short review of some of the most noticeable defects may afford useful lessons to the novice. Many hall-stands are too narrow from back to front, so that they have little stability. This must be so if people want the orthodox piece of furniture. Then the coats are often in the way of umbrellas, &c., and the glass is not infrequently in such a position that a rail comes just about the height of a tall man's head, and if there is a box with a lifting lid, everything must be removed before the interior can be got at. A well-designed—apart from merely pretty—hall-stand should stand so firmly that it does not fall forward when full of coats; these and umbrellas should be out of each other's way, glass at such a height that it can be used without stooping, and drawer or box easily accessible. If these features are wanting, merely decorative detail cannot make a good hall-stand.

In the same way every piece of furniture should fulfil its ostensible purpose, for decoration and good workmanship are poor substitutes for convenience and utility.

Some measurements seem purely arbitrary, and to be regulated entirely by custom, which alters. Thus sideboards a century ago were much deeper from back to front than now, and the back, if any, was low. At the same time, there are sizes which are generally recognised as being customary. The lower or carcase part of a good 6-ft. sideboard will stand about 3 ft. high, though it may vary a little either way.

From all this it will be seen that the designer is confined by little beyond convenience and custom in arranging the size of furniture, though very frequently the cost has to be taken into consideration. This point need not be insisted on here more than by saying it is an important factor in deciding the substance of the material. If this is thick enough and strong enough nothing more is absolutely necessary, though custom and appearance have to be consulted and consequently taken into consideration. The sideboard top of $\frac{1}{2}$ in. thickness is all that utility demands, but good appearance, or what we consider such, otherwise custom, requires it to look more, consequently it has a thick moulded edge given to it by lining.

The first principles of design in furniture the beginner

should note are that the thing should be fit for its purpose, that the construction should be sound, and that purely decorative details should only be added to good construction. To make these ornamental details the first consideration would be altogether wrong. Ornament should be added to construction, instead of the reverse being the motive of the designer.

It seems almost needless to remark that some knowledge of drawing or facility with pencil is essential to those who would make their own designs, though to make these properly requires special knowledge and training beyond that which is given in ordinary schools, even if they be of art in 'connexion with South Kensington.' Technical details have to be understood. These are of more importance to the worker than power to produce a finished drawing or of a knowledge of perspective. Both these may be of advantage, but a nice-looking drawing is of very little use in the workshop if incorrect.

Any one who can use a T square and the other essentials to mechanical drawing can make a small drawing sufficiently good for workshop purposes, provided he has some knowledge of construction. The small drawing is simply the sketch from which the working drawing is set out, and may be either to scale or drawn in rough perspective. The former is the more accurate, though perhaps to the ordinary observer the latter may convey a better idea of the thing intended to to be represented. Fig. 159 shows a small overmantel so treated, and Figs. 160 and 161 represent the same thing drawn to scale, both front and end elevations being shown. For workshop purposes, if half only of the front were drawn it would be sufficient, and the sketch might be of the roughest. As a matter of fact they generally are.

To the actual workman the working or full-sized

214 *DRAWING AND DESIGNING.*

drawing is of far more importance than the miniature design, for it shows him the thickness and size of each

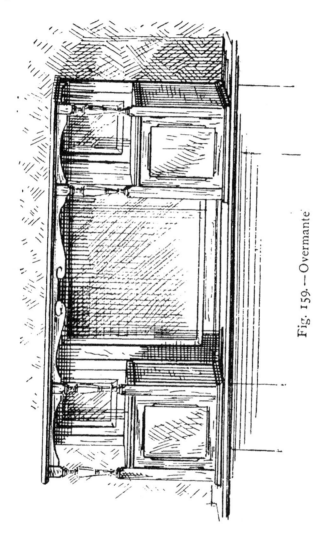

Fig. 159.—Overmantel

piece of wood, as well as of the whole, and in some cases the details of construction, joints, &c. These are not always necessary, and a working drawing is often a very rough and fragmentary affair, not at all the thing a drawing-master would approve of. If it shows sufficient

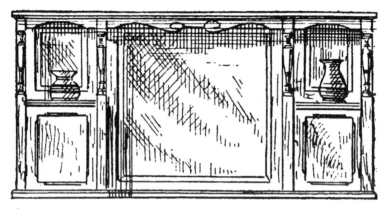

Fig. 160.—Front Elevation of Overmantel.

to guide the worker nothing more is needed, though it may look better if more highly finished. As might be expected in actual practice, the extent of detail shown in working drawings varies considerably. In some there is little more than a full-sized elevation, while in others every joint almost is shown. In the vast majority of cases the latter is quite unnecessary, for the maker knows which method of construction to employ without having it before him in the drawing. Too much detail, that is, more than is required, unnecessarily complicates the drawing, so that the worker may be advised to make these as simple as they can be consistently with showing what is wanted.

Fig. 161.—End Elevation of Overmantel.

Working drawings, it may be remarked, are often dispensed with altogether, but the novice certainly should not attempt to begin any job without one.

With a drawing he has something to guide him and to refer to when in doubt. The time spent in preparing it will be more than saved afterwards, not to speak of the smaller liability of wasting material by cutting it wrongly.

The drawing may be either to scale or full-sized. The latter is preferable. The scale means that all the parts are proportionate, but are drawn in small size: for example, 1 in. scale means that each inch of the drawing represents a foot of actual measurement, so that 1 in. of this in the drawing is $\frac{1}{12}$ in. The scale may be anything, $1\frac{1}{2}$ in., 3 ins., &c., which simply means that each of these is equal to a foot. Drawings on a small scale naturally cannot be made so accurately as those which are of full size.

The construction of working drawing does not require a knowledge of ordinary drawing, though, of course, this can be no disadvantage to the draughtsman. Many, however, can make good working drawings who are unable to draw at all in the ordinary sense, so the reader, if he is among the latter class, need not fear that he will be unable to do what is wanted. If he can rule lines with the square very little more will be wanted except head work, for every drawing of the kind wants thinking about. It cannot be done mechanically.

First of all the front elevation should be drawn, that is, the outlines of each part represented flat, no shading nor perspective being required. Then, if necessary, an end elevation is drawn in the same way.

To show the thickness of various parts, sections are drawn on them, and are indicated by transverse lines. A sectional drawing means that the thing represented is supposed to have been cut in two showing the edges of all parts.

A plan is a drawing of the lines of a thing from above. Thus a plan of a table, 3 ft. by 2 ft., is nothing more than four lines forming an oblong of this size.

On it can be indicated the position and thickness of the legs and of the framing, but these can be shown in another way, viz., by drawing an end elevation and a front one. These, of course, show the top too, so that there is no occasion for all the drawings to be made. As a matter of fact, a drawing would seldom be used for a small plain table. It has only been named as a simple example.

A working drawing which may come before one is seldom, indeed never, comprehensible at first, unless of a very simple character. Its lines require studying. It must be read as it were before its intention becomes clear and the bearings of the relative parts to each other are understood. Some drawings require a great deal of study, so the novice should not be discouraged if he cannot make out their drift at first. It is very often the custom to supplement the drawings with written explanations, and though words are not parts of a drawing, none but a pedant could object if their use elucidates anything. I want my readers if possible to get away from the notion that there is either any mystery in making or understanding a working drawing, or that a number of rules could help them. Common sense is the best guide, after the elementary principles which have been given; and probably it would be as difficult to get several men to make a working drawing exactly alike as to get them to write a letter in the same words. The sense would be the same, but not the means by which the intention is conveyed. However much the drawings of any piece of furniture might differ in appearance, if made by competent draughtsmen and worked from by equally competent cabinet-makers, the articles made would resemble each other. If the constructive joints were not shown they would probably differ, but so far as external appearance the things would be alike.

Now, after this a rendering of the overmantel named (p. 213) in working-drawing style will be of assistance to the beginner. As both sides are alike only one half need be shown of the front.

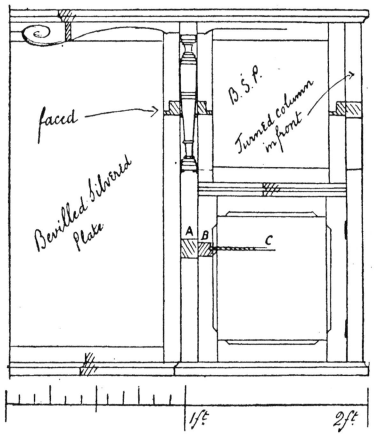

Fig. 162.—Working Drawing and Scale of Overmantel.

Necessarily it cannot be given here (Fig. 162) full-sized, but is drawn to the scale which accompanies it.

First of all the principal lines are drawn, and then the sections put in. Analysing the drawing, we find that the thickness and length of the top shelf, the top and bottom of the cupboard, are got. Their width is seen from the end elevation (Fig. 163). The thickness of the columns is seen by section A, which shows them to be square. Following along to the right we find that the door frame B is not so thick as the square, and that it is set back a trifle, for were it flush with the front of the column the line would go straight. Further, we find that the door frame is rabbeted, and that the panel C is held in by beads. Now there is no use in extending the sectional drawing right across as it already shows all that is necessary. The *thickness* of the ends is shown at D, and might have been drawn against A except for the additional clearness of its present position. The width of the end is got from Fig. 163. The section of the back is sufficiently indicated and explained by words as well as drawing. Now this does not profess to be a perfect drawing, but it is as much as would be required, and gives a better idea of a working drawing, as practically used, than a more elaborately worked out one would do. Any cabinet-maker can see what is wanted, and more than many would require is shown.

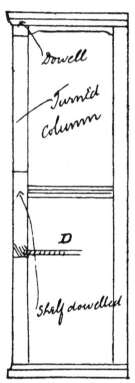

Fig. 163.—End Elevation.

To show the construction equally clearly by a totally different arrangement of drawings would be quite possible, and might be interesting, but would be of no real utility, as the same principles would be found embodied in all.

The setting out or working drawing may be made on paper, but it is more convenient to do so on a piece of board unless the article of furniture is very large. It will rarely be found necessary to draw more than half of anything, if so much ; for, as in the case of the overmantel, words and figures may be used to show the length of the central space. A thin pine board, or several of them joined together to make the necessary width, is as useful as anything. The drawing can be planed off afterwards, and the same wood used either for another drawing or for working up. A pencil will be used to draw with, and if it does not mark distinctly enough rub the wood over with chalk beforehand.

Nothing more need be said about drawing except to recommend the reader to endeavour to understand any he may meet with even if he does not intend to make up the article shown, for he will by so doing gather much information which cannot be acquired otherwise.

CHAPTER XV.

VENEERING.

*O*bjections considered—Burr Veneers—Saw-cut and Knife-cut Veneers—Laying with Caul—Wooden Cauls—Metal Cauls—Care and Preparation of Veneers—Preparation of Wood for Veneering on—Light coloured Veneers—Cleaning up Veneered Work—Laying with Hammer Veneering Hammer—Blisters—Veneering on End Grain—Inlaid Veneers—Veneering Curved Surfaces.

VENEERING is hardly practised now to the same extent that it was at one time, but it still forms a very important part of the cabinet-maker's work, and no one could be considered a proficient unless able to do it thoroughly. The simple notion that any one can stick two pieces of wood together with glue is a very crude idea of veneering, for there are many details which must be regarded if the work is to be thorough. Merely to glue the wood and veneer and then stick them together anyhow is by no means sufficient, for such work could not stand.

Veneering, in many quarters, is regarded with suspicion, but when properly done there is nothing objectionable about it. It is sometimes assumed, but not by cabinet-makers themselves, that the principal object of veneering work is on account of its supposed cheapness. One writer goes the length of saying that it is never worth while to veneer with mahogany, as the solid wood can be used at very little more cost. Now, if he had known what he was writing about he would have been aware that it is quite possible to have anything made of solid mahogany at considerably less than it would cost to have it veneered. A really fine veneer may be, and often is, more valuable than thicker stuff with less figure in it. Then, in addition to the cost of materials, there is the labour to be reckoned for, so that it will be seen that veneered work does not necessarily mean cheap work. When a valuable veneer is used it is seldom laid on poor wood as a foundation, for no cabinet-maker would think of using such. A choice Spanish veneer, for instance, would be laid on mahogany, and it is, in such a case, not wrong to consider anything so made as being of solid mahogany, for it really is so. The solid is covered with a choice thin wood, and no one with the smallest knowledge of the subject would expect to find the wood throughout the same as the veneer. Many of the most beautiful woods cannot be used except in the form of veneer, so that if we are to listen to those who object to veneering these would have to be entirely neglected. Among the objections raised to veneering, one sometimes hears that it is used to conceal defective workmanship, that joints covered with veneer are weak, and so on. As a matter of fact, the work is done equally well by all respectable cabinet-makers whether the work is veneered or not. Veneer may be used to conceal bad workmanship, but only those who would 'scamp' in other forms would resort to such trickery. It is by no means the rule to find that veneered furniture is put together in any inferior way than would be practised in making things of similar quality without veneer.

Of course, veneer may be used for the purpose of making cheap furniture, and often is employed in preference to solid wood for the express purpose of reducing cost of production. With it a thing made principally of pine may be as handsome in appearance as if it were made of mahogany or walnut, and for practical purposes is sufficiently useful. To say that a comparatively choice wood should not be used on a cheaper and less beautiful one seems absurd. Veneer certainly allows many to surround themselves with good-looking furniture instead of plain pine. Naturally it costs more than if made in the latter solely, but is not by any means so expensive as solid (say) mahogany or oak, unless perhaps these are of the commonest description. In any case, veneering may be regarded as a ready means of adorning furniture at a comparatively cheap rate, and so placing beautiful objects within the reach of many who could not otherwise afford them, and surely there can be nothing wrong in doing this. Mind, I am not arguing that beauty, or, perhaps I should rather say, appearance, is desirable at the expense of sound construction, only I wish to protest against the assumption that this is not consistent with or commonly found in connexion with veneer. That this is sometimes, indeed often, found improperly laid, is not to be denied, but then it is the workmanship which is to be blamed, and not the entire principle of veneering indiscriminately. On the whole the cabinet-maker need have no hesitation in using veneer whenever it can be employed with advantage, taking the same care over the construction as if only solid wood were used, and laying the veneers properly. The wood serving as the foundation, whatever its kind, must be good and sound, free from knots and shakes or cracks, and, in fact, just such as would be used if it were not to be veneered, except in the matter of figuring.

As woods which cannot be used in the solid have been referred to, it may be said they include all of the 'burr' class. In these the markings are very elaborate, and the kind most familiar is probably burr walnut. Amboyna is another of the same sort, of a very beautifully marked deep yellow or light brown colour. Thuja is somewhat similar, but deeper in tone, and with strongly marked eyes or knots. Pollard oak of the choicest kind is also usable only in veneers. Roughly speaking, most of the furniture woods can be had in veneer form as well as solid, the exceptions being, of course, the cheapest kinds, such as pine, whitewood, &c. Not only the choicer varieties of each kind are cut into veneers, but the plain ones as well for common furniture. Thus we find veneers of ordinary American walnut.

Veneers are of two kinds, known as saw-cut and knife-cut. The latter are much thinner than the former, and are generally plain in figure, so that they are much lower in price. It is rare to find choice material in the form of knife-cut, so that when anything good is to be made, the saw-cut should be preferred. Even this is of no great thickness, and is to be had in a larger range of quality than the other. It is, perhaps, rather too much to say that knife-cut should never be used, but it certainly cannot be recommended for general purposes. Unless otherwise specified, it is generally understood that veneers are of the sawn variety. As it has been said that knife-cut is not to be recommended, it may be as well to add that this remark does not apply to burr veneer.

Like solid wood, most veneers are reckoned by the superficial foot, burrs being generally sold at so much each.

Veneers are laid, that is, glued to the solid backing, by two methods, viz., with the caul or with the hammer.

Caul laying is the better, though more troublesome, of the two processes, and may be described first. With it the veneer and ground are kept in close contact by pressure till the glue has hardened, so that they are as firmly joined as possible if the work is done properly. The cauls are simply pieces of wood or metal which fit exactly to the veneered surface. It is rarely that the cabinet-maker requires to use any but flat cauls, as it is seldom that a curved surface has to be veneered. If he should meet with any, if he understands flat work, he will have no difficulty in knowing how to act. In order, therefore, to make the explanation of the process as clear as possible, it may for the present be assumed that only flat surfaces have to be veneered.

Wooden cauls have been generally displaced in well-appointed shops, where much veneering is done, in favour of metal, but for occasional use they do very well, the principal advantage of metal being that it is more durable. There are, however, disadvantages attending its use, and many consider that the wooden caul is preferable, so that the novice or amateur has no occasion to use anything else. The caul is merely a piece of wood of, say, an inch thick, of such a size as will cover the veneer. Pine does as well as anything, and a nice clean piece which is not likely to twist should be selected. It may be worth while noting that wood which has been used for a caul may be employed for working up afterwards. Its use as a caul will, at any rate, have made it thoroughly dry. Generally, however, the cauls are kept as such, but on occasion any piece of wood may act as a temporary caul without being destroyed for other purposes.

For metal cauls sheets of zinc are generally used, as being the most suitable. The thickness is not of importance, and for ordinary purposes may be anything from $\frac{1}{4}$ in. to $\frac{1}{8}$ in., or even less, as it is advisable to use it in conjunction with a wooden backing or caul in most cases. The advantage of metal over wood consists principally in its durability, as it is not so apt to be destroyed by careless heating; but with ordinary care there is no reason why a wooden caul should be burnt. Perhaps the chief objection to a metal caul, especially when used by a novice, is that it may be made too hot for the work without being destroyed. A wooden one may be made too hot, but in keeping it from being charred the beginner is not likely to make it so. In competent hands the work can be done equally well with either wood or metal, and extreme nicety in the degree of heat is not important. But this will be referred to later on.

The treatment or preparation of veneers is simple; in

fact, beyond having them thoroughly dry before beginning to lay them, little need be said about most of them. If they are kept in stock, to prevent them splitting when moved about, the ends should be covered with paper, or cotton, or something of the kind glued over them. Burr veneers being very much twisted, require flattening out before they can be laid. The best way to treat them is to damp them till they are sufficiently pliable, and then put them between two heated boards or cauls, pressure being exerted by hand-screws. When dry, they will be flat, but before they can be laid the holes and imperfections must be filled up, and to do this properly is a troublesome job sometimes. Burr walnut veneers, it may be well to explain, are full of irregular holes and cracks, and in their original state have very little indication of their beautiful appearance when laid and polished. The holes are of such a kind that they cannot be filled by simply stopping them up, for it is generally necessary to cut away some of the surrounding wood and let another piece in. To ensure an accurate fit, two pieces may be sawn through together with an ordinary small fret-saw, and in doing this care should be taken that the inlaid piece matches as nearly as possible to the figure of the surrounding wood. By sawing an irregularly shaped piece to correspond with the markings of the wood, the joint will in the finished veneer scarcely be observable. When the work has been really well done, it is often almost impossible, even with close examination, to discover the pieces which have been let in. I suppose it is hardly necessary to say that an entire sheet need not to be cut up to supply the small inlet pieces, as small bits of odds and ends will do very well for them. As they are cut, fit them in to the holes, and glue pieces of paper over to hold them in. If the veneers are choice, no pains should be spared over this part of the work, as nothing looks worse than to see a fine veneered surface marred

VENEERING.

by pieces let in so conspicuously that one cannot help noticing the inlaid bits.

Any veneer which may be twisted and require flattening may be treated between cauls after having been previously damped. With rosewood, however, it is better to warm it instead of damping. It should be held near the fire till the oil or resin exudes, and then placed under pressure till cold. Even if already sufficiently flat, it is often considered necessary to heat it, in order to get rid of some of the resin it contains. This applies especially to those kinds which contain much resin.

Whenever veneer has been damped, it is necessary to be very careful to see that it is thoroughly dry before laying. As much care should be taken with all veneers in this respect as with solid wood.

Before the veneer is laid, any saw marks should be removed with the toothing plane, which at the same time roughens the surface slightly and affords a good hold for the glue. The surface of the groundwork should be prepared in the same way, for if left unplaned, the saw marks will be observable through the veneer. In fact, the wood should be prepared as carefully as if it were not to be veneered, and also be toothed. Knife-cut veneers having no saw marks, may be laid just as they are; indeed, they are so thin that they would not stand toothing.

The veneer should be cut slightly larger than the wood it is to be placed on, leaving the rough edges to be cleaned up afterwards. With regard to the ground work something must be said, for there are several particulars to be noted if the wood is of any considerable width, such as panels, carcase ends, and so on. At the same time it must be remarked that in details considerable variety of practice is found. Few men who have done much veneering work but have some pet method of their own, and details will be found to differ. Modern practice tends towards increasing simplicity in place of some of the more complicated and cumbersome arrangements which formerly prevailed.

The side of the wood on which the veneer is laid is a matter of some importance. If laid on one side, the veneered surface will become hollow, and if on the other the reverse. That is to say, if the panel or board does not remain flat, as it may do, the tendency will be in these directions. Now in the majority of cases a hollow surface is much more objectionable than one slightly rounded, so that it is best to lay veneers on the heart side of the wood. I may even go further, and say that the contrary should never be done unless from quite exceptional reasons. As novices may not understand which is the 'heart' side of a board, it may be explained as being that which was nearest the centre of the tree trunk. It can easily be detected by noticing the direction of the annular markings on the end grain. If they are not visible owing to the roughness left by the saw the end can be planed, and they will then be clearly seen. To prevent mistakes, Fig. 164 is given. It represents the end of a board, and shows the upper surface as being the heart side, the one on which the veneer ought to be laid. I am particular about this, as the novice may meet with instructions telling him to do just the contrary, and be slightly perplexed thereby. Well, as I have suggested elsewhere, let him follow the practice of good workmen, and that is to lay the veneer on the heart side. Some cabinet-makers may do the reverse, but they will be rather difficult to find, and when found will probably be indifferent specimens.

Fig. 164.—Section showing Heart Side uppermost.

The back of the wood should be slightly swelled or rounded by damping, and about this it is impossible to give minute directions, for so much depends on circumstances, size, thickness, &c., that very little can be said about it. It must also be observed that the degree of moisture or swelling given by different workers varies considerably. At one time it was considered advisable to soak the wood thoroughly for a day or two—it may be so still, by some, for aught I know—before veneering. It is, however, by some generally deemed sufficient, and rightly so, to simply damp and swell the wood on the back for a few hours beforehand. This may be managed by wiping the wood over from time to time with a damp rag or sponge, or, and perhaps more conveniently, laying some damp sawdust on it. The only objection to this latter course is that if left too long the sawdust may act more than was intended. However, experience alone can determine what is required in each instance.

The wood being ready is laid on the bench, hollow or heart side up. Strong, *i.e.*, good fresh hot glue, is then rubbed on it, not on the veneer, which is then laid without loss of time. In the meantime the caul or cauls should have been made hot, and are then placed above the veneer. To prevent this shifting, one or two veneer pins may be driven in where necessary, but it is as well to avoid using them excessively. If carefully inserted and bent over they can, however, easily be withdrawn afterwards, and need not injure the veneer, as the marks will not be noticeable. In the case of panels they can be driven in by the edges, which will be concealed behind the rabbet. If the wood is only thin, it should be laid between cauls to keep the veneer close to the upper one. Pressure is then applied by means of the hand-screws, which should be liberally used. The heat from the caul will strike through the veneer and melt the glue, causing the surplus to run out at the edges and the surfaces to come into the closest contact. From this it will be seen that it is more important that the caul should be slightly rounded if anything rather than flat, so that the glue may be pressed from the middle outwards to the edges. With a hollow caul of course the tendency is to drive the glue to the centre, and this must be avoided. It is also necessary that the caul should be uniformly heated, so that the glue is equally softened all over. If made too hot the glue, instead of being merely melted, might be burnt or spoiled, and the veneer be injured. With wooden cauls this risk is a very small one, and even with metal is not a very serious one, as if made too hot the plate could not be comfortably moved with the naked hands. The novice with these cautions is hardly likely to err except by carelessness. Metal cauls, unless of exceptional thickness, should have a board laid over them to equalise the pressure on the veneer.

Some of the glue will be forced into and perhaps through the veneer. To prevent the caul sticking it should, if metal, be slightly greased, and a sheet of paper be laid between it and the veneer. With a wooden caul one or two sheets of paper will be sufficient. These precautions are specially necessary with a porous open veneer like burr walnut, but may often be omitted with those of closer texture.

As the glue in striking through is apt to discolour very light veneers, a little more care is necessary with these. A colourless glue may be used, but as this is not always convenient the ordinary kind will do perfectly well if mixed with white, such as powdered chalk, whiting, &c. Whatever the material, it should be finely ground and free from lumps. Glue so prepared is used in the ordinary way, or the veneer may be sized with a little of it thinned down sufficiently. When this latter is done the veneer should be allowed to dry before laying. Another method is simply to rub dry chalk on the veneer and on the board, so that the mixture is formed while the glue is being rubbed in. With light veneers it is advisable to use the glue as thick as it

VENEERING. 231

conveniently can be, and to lay it on thinly. As some misapprehension may arise from light veneers being mentioned, it may be said that with oak, ash, and such ordinary woods it is seldom necessary to use anything but good ordinary glue, only the finer and whiter veneers requiring chalk, &c. Of course, even oak and ash, if they are to be kept as white as possible, may be laid with whitened glue, which can be used without injury whenever it may seem desirable.

The cauls should be left on till cold, and even then there need be no hurry in taking them off, as they cause no damage by being left on. When several pieces of the same size, such as drawer fronts, are being veneered, two of them may be laid back to back and a caul applied to each veneered surface, or they may be laid together with a caul between them.

When taken out of the caul the work should be laid on one side for a time to allow the glue to thoroughly harden before cleaning up. It is best to keep the veneered sides from the air either by placing them next a wall or by putting them together in pairs when there is more than one piece. To work on the veneers, or as it would be said in the workshop, to clean them up before the glue is thoroughly hard, would be to weaken the hold, and very likely cause blisters. These will be more fully referred to in due course. It must be noted that the heat caused by the use of the scraper and papering up is often sufficient to soften the glue so long as this retains any moisture, and if this is done the veneer is almost sure to rise. It may almost be said that to clean up veneers too soon is to ruin the work and render previous care futile. How long the veneers should remain untouched before cleaning up must depend on circumstances, and any cabinet-maker's advice would probably be to leave them alone as long as possible. There is no chance of justice being done to the work if they are cleaned up within two or three days, and if they are left for as many weeks it will not suffer.

If the veneer is thick enough the smoothing plane may be used if necessary, but it should, of course, be set as finely as possible, and very little be taken off with it, otherwise the scraper and glass paper alone must be used. When veneers are laid with the hammer cauls are not used, the contact being effected by simply passing over the work with pressure. Some cabinet-makers are of opinion that this method is as good as the other, or at any rate sufficiently so, but as a matter of fact the work is seldom so reliable unless for small pieces, and even then the caul is better when it can be used with convenience. In this respect, a good deal depends on the discretion of the worker, and there is no doubt that many are very successful in veneering even large panels with the hammer. The novice, however, cannot be recommended to use this for the purpose. All that can be said is that he may do so, and he must not be surprised if, occasionally, the work does not stand. With knife-cut veneer, the difficulty of doing the work properly with the hammer is less than with the thicker saw-cut. If I may venture to do so without being understood to fix arbitrary limits, I would say that the caul should be used with saw-cut, and that the hammer should be kept for knife-cut veneers, especially when large pieces are being laid. There is, however, nothing but experience to show when it is a matter of indifference whether caul or hammer be employed. The latter is often the easier and quicker for small work, such as rails and stiles of door frames, but it is really impossible to say exactly, and much must be left to the judgment of the worker. The broad end of the ordinary hammer may be used for narrow strips, but for general purposes a special veneering hammer is advisable, and will be

VENEERING.

more convenient. It can easily be made by or for the user, and is not generally on sale at tool-shops.

Its construction will be understood from the accompanying illustration, Fig. 165, and the following description:—The head consists of a piece of strong wood say 1 in. thick; into the lower edge of this a piece of iron or steel about $\frac{1}{8}$ in. thick is firmly secured, with, say, an inch projecting below. A wooden handle completes the contrivance, which is really more of a squeegee than a hammer. The size is comparatively unimportant, and depends more on the fancy of the user than on anything else, but to give some general idea of good useful dimensions the following may be stated:—Length of handle from 12 to 18 ins., head 6 to 9 ins. long and 4 to 6 ins. wide. The iron must be of such a thickness that it does not bend under the pressure, and should be straight across, with a slightly rounded edge and corners to prevent the veneer being scraped or scratched, instead of being merely rubbed. From these remarks it will be seen that the tool need only be a rough one. Indeed, fine finish would be thrown away, as it soon gets dirty with use.

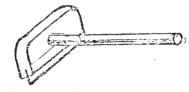

Fig. 165.—Veneering Hammer.

When veneering with the hammer, the wood is not swelled on the back and the glue is simply rubbed on the face side. When this has been done, the veneer should be laid and pressed down as quickly as possible, so that the glue may not have cooled more than can be helped. As this is of importance when laying with the hammer, everything required should be at hand. A rag or sponge, some clean hot water (that from the outer glue-pot ought to do very well), and a heated flat iron such as used for laundry purposes should be ready. When the veneer has been hastily pressed down with the hands, the hammer will come into use to complete the work by forcing out the air and surplus glue. First go over the veneer with the sponge or rag moderately wrung out with the hot water as quickly as possible. Then take the hammer with the handle in the right hand, the left pressing down on the head, and with the iron on the veneer go over the whole surface by means of a series of zigzag movements as suggested by Fig. 166.

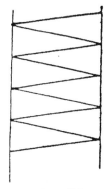

Fig. 166.—Diagram showing course of Veneering Hammer.

Naturally the centre of the panel should be treated first, so as to gradually work the air and glue towards the edges. In order to let the hammer work freely, the veneer should be kept tolerably wet, as mere dampness would not do; on the other hand, sloppiness should be avoided, and the water should be hot. It is not a bad plan to have a little glue in the water, though by wiping the edges with the sponge or rag enough will be got, and it is by no means necessary to mix glue and water specially. Before the veneer can be properly laid, especially if the panel is a large one, the glue will in all probability have become so set that it does not flow freely. To remelt where necessary go over the veneer with the warmed flat iron, which, be it noted, should not be too hot, and then continue rubbing with the hammer. As the work progresses blisters, which are really formed by air bubbles under the veneer, should be watched for. If removed at once, that is, by working them out to the edge of the wood while the glue is hot, there is not much difficulty with them, but if left till the glue has become hard it is often troublesome to get rid of them. To leave them in finished work would never do. When the veneer seems, so far as can be seen, properly laid, go over it carefully for the express purpose of discovering

VENEERING.

any blisters. These are often difficult to discover with the eye alone, but may easily be detected by lightly tapping on the veneer with the hammer handle; the hollow sound will betray them. It may be possible to lay the veneer at such places by simply reheating the glue and using pressure, but to allow the air to escape it will be necessary to prick a hole or make a short cut with a sharp chisel. If done with the grain of the wood the mark will not be noticeable. To heat the glue locally a warm hammer head may be large enough or the bottom of the inner glue pot answers very well. In some cases the blisters will be found to be caused not so much by air bubbles alone as by the glue having been all run out. It is then necessary to put some more under the unstuck parts of the veneer, and it may generally be managed by cutting through sufficiently to get a little in. If this cannot be done by a simple plain cut it may be necessary to cut so that a portion of the veneer can be partially raised. With some woods this is easy, but with others great care must be taken in order that the marks may not show afterwards. Blisters, it may be remarked, are not so likely to occur when the caul is used as with the hammer, but whenever they are found they may be laid as suggested. If there are many it is an evidence that the work has been badly managed, for with a fair amount of skill and ordinary care they will seldom occur. When a very soft or porous wood forms the ground it is sometimes considered advisable to prepare it by sizing over with weak glue, but if attention is paid to the consistency of the glue used to stick the veneer down with this is very rarely necessary.

Whenever, as occasionally happens, it is necessary to fasten veneer on end-grain wood, the glue must be well rubbed into this and allowed to dry. By this means the pores of the wood are closed, and the glue which is afterwards applied does not soak in too much. Another plan sometimes adopted when time is an object is to rub the glue in as before, and then, instead of waiting for it to dry naturally, pass a hot iron over. This sears the glue and chokes the wood up as before, but it is hardly such a good plan.

Inlaid and burr veneers cannot be laid with the hammer, but must be done with the hot caul. The paper with which they are either entirely or partially covered should not be removed till they are cleaned up after laying.

It is occasionally necessary to veneer on both sides of a panel; in this case the wood does not require swelling on the back beforehand, as the hot glue being applied to both surfaces renders this superfluous.

Unless a large number of curved pieces requiring correspondingly shaped cauls have to be veneered, it is often not worth while to make a wooden or metal caul, even when the veneer is to be laid by this method. For ordinary purposes what is to all intents a flexible caul may be used with greater convenience. If the worker will remember that what is wanted is equable pressure with heat, he will be helped over many a difficulty, for in veneering, as in other operations, it is absolutely impossible to give specific directions for every case which may arise. One of the most useful substitutes for a wooden or metal caul consists of a stout ticking or canvas bag containing sand. This will adapt itself to any curved work to which it may be applied, and is easily heated to the requisite degree. By placing on it, after it has been put in position, pieces of board which force the sand caul as closely as possible to the veneer, and then applying the hand-screws, any required pressure can be got and maintained.

Various other expedients of a similar character are frequently adopted, but few of them are of such general utility as that mentioned, for with it it will rarely happen that an odd piece of curved work cannot be veneered with ease, and almost as speedily as with a wooden caul.

When the veneer has to adapt itself to a very sharp curve it is sometimes advisable to back it with a piece of calico or canvas, in order to prevent its breaking or splitting.

It is comparatively rarely that cylindrical work has to be veneered, still, as it occasionally happens, it may be well to say that a method which is as good as any for it is, after cutting, glueing, and laying the veneer, to bind it tightly round with ordinary chair webbing. If this is then slightly damped it will draw tight, answering to some extent both as a caul and instead of hand-screws. The work should be placed near a fire, in order that the adhesion may be perfect.

When laying veneers on the rails and stiles of door-frames, which, by the way, is generally done with the hammer, it is advisable to glue strips of paper over the joints and leave them till the glue has become thoroughly set. If this precaution be neglected the joints are almost certain to open as the veneer dries. It may be noted that no harm can result by glueing paper over all joints in veneer. This may not always be necessary, but on the principle of prevention being better than cure many good workers make a practice of never omitting it.

It seems hardly necessary to point out that on door-rails and similar parts the veneer should be laid with its grain running in the same direction as that of the ground wood. This is now generally done, but till very recent years it was no uncommon occurrence to find that the opposite course was adopted. The grain of the veneer, instead of being horizontal or coincident with that of the rails, was perpendicular and parallel with the stiles. This arrangement gives the work an unnatural look, and would not now be tolerated in good work in this country. Cross-grained rails give an impression of weakness, and as a rule the worker should lay veneers with their grain running in the same direction as that of the ground wood. In connexion with this it may be said that almost the only parts which are now veneered differently are plinths and similar parts of round-cornered work, such as chests of drawers, wardrobes, &c. These are often found with the veneer grain perpendicular instead of horizontal; but no doubt even this peculiarity will in due time go the way of other barbarisms in furniture of the early Victorian period.

CHAPTER XVI.

CABINET BRASS-WORK.

Till Locks — Cut Cupboard Locks — Box Locks — Desk Locks — Straight Cupboard Locks — Wardrobe Locks — Nettlefold's Piano Lock — Spring Catches — Flush Bolts — Socket Castors — Screw Castors — Pin Castors — Castor Rims — Dining-table and Pivot Castors — Iron-plate Castors — Ball Castors — Wright's Ball Castor — Paw Castors — Butt Hinges — Back-flap Hinges — Card-table Hinges — Desk Hinges — Screen Hinges Centre Hinges — Piano Hinges — Hinge Plates — Escutcheon Plates and Thread Escutcheons — Brass Handles.

To treat of cabinet brass-work in anything like an exhaustive manner in the present volume is, of course, out of the question; but to entirely ignore the more important articles would hardly be considerate towards the novice. Those things which are of general utility therefore, will receive attention in the present chapter, and will at any rate serve to show something of the extent to which the cabinet-maker is catered for by the cabinet brass-founder. Somehow or other the subject has almost escaped notice in handbooks purporting to be guides to the construction of furniture, or where it has been referred to in technical works the writers have concerned themselves more with the things used by builders and joiners than with those of the cabinet-maker. At present, of course, only those things which concern him will be mentioned.

Among the most prominent of these are locks, of which there are many varieties, not only in quality but in shape and general arrangement, to enable them to be used in various situations. With the peculiarities of construction and the differences between the various kinds, so far as the mechanism of the interior is concerned, it will be unnecessary to waste space. The quality is more a matter of price than anything else, whether the lock be lever, tumbler, or one of the somewhat numerous patents. For ordinary purposes, lever locks are generally more popular than other forms, but the user may safely consult his own ideas on the kind he gets.

The *till* lock is for drawers and similar parts of work. In it the bolt shoots upwards when the key is turned, catching in the fixed wood above, and so prevents the part to which it is attached being withdrawn. When fixing this or other locks which have to be let into the wood, the space it is to occupy should be carefully marked out with gauge and square, the waste wood being removed with a chisel. The exact place in which to cut the hole for the bolt to shoot into may be somewhat troublesome for the novice unless he knows how to find it. All difficulties, however, will disappear if he adopts the following expedient: After fixing the lock, smear the top of the bolt with a little colour of any kind —gas-black and glue, or anything of the sort—close the drawer and then turn the key as far as it will go. The bolt will then shoot against the bearer and leave an imprint showing the exact size and position of the space to be cut out. This may be done with an ordinary chisel, although there are bent ones made for the purpose. These, however, are not necessary in ordinary work. The position of keyholes must be accurately gauged, and they should be carefully cut, as nothing looks worse than a slovenly hole. The wider, or round part, for the shank of the key, should be bored and the rest cut out with a fine saw or chisel, as may be most convenient, and finished by filing, when necessary. For ordinary drawer work the most useful size is $2\frac{3}{4}$ inches long, with a distance to pin, *i.e.*, space between top of lock through which bolt shoots and the pin on which key turns, $\frac{3}{4}$ in. and $\frac{7}{8}$ in.

It may be mentioned that in some locks there are no pins, the end of the key fitting into a hole bored through the outer plate of the lock. This form is sometimes considered to have advantages over the ordinary pipe key, inasmuch as it cannot get choked up with dirt and there is no lock-pin to be broken or displaced.

Cut cupboard locks are exactly similar to drawer locks in general principle and arrangement, the only difference being that as they are fixed in door-frames the bolts shoot at right angles to the keyholes. Were it not that a keyhole cut across a door-stile looks awkward, the drawer lock might be used instead of the cut cupboard lock, and *vice versa* in the case of drawers. Some locks having keyholes cut both ways, may be used either for cupboards or drawers. Cupboard locks are made right or left handed, *i.e.*, for the bolt to shoot either to the right or left, so that by the exercise of a little discretion the cabinet-maker should never be under the necessity of making a keyhole upside down. In all cases the distance to pin should be considered, so that the keyhole may come fairly in the centre of the stile, though absolute exactitude in this respect can hardly be considered necessary, and is not always possible.

Box locks are, as their names indicates, used for boxes and parts where they can be fixed under similar conditions. In them the bolt, or rather bolts, are altogether concealed within the lock, hooking into links which sink into holes in the edges and forming parts of the plates which are always supplied with such locks. To find the position for fixing the plates, which of course should be sunk flush, it is only necessary to put the plate on the lock after this has been secured and shut down the lid so that this is marked.

R

When cut with keyholes in three ways, these locks may be used, as they often are, for right or left handed locks for sideboard pedestal or other doors, when these cover the ends, as well as for their ostensible purpose. When used for pedestals the locks, of course, are fitted to the ends and the plates on the doors, which—to simplify explanation—may then be considered as lids fitted perpendicularly.

Desk locks are similar in principle to those last-named, which, in office and other large desks, are often, indeed generally, used instead of the former. In the form ordinarily used on small fancy or portable desks, and specially entitled to the designation of desk lock, the hook-shaped bolts project, fitting in and hooking to plates with holes through them. It is, therefore, necessary not only to sink the plate, but to cut spaces for the bolts. A variety of lock with hook bolts is so made that these lie flush when not in use.

Straight cupboard locks are perhaps the most easily fitted of any kind, as instead of being cut into the wood they are simply fastened on the back of the door. Their position is in fact the reverse of the cut cupboard lock, as what in this is the external plate is against the wood, the remainder projecting. From this it may be imagined that the straight cupboard lock is not a desirable one when neat appearance is a primary consideration, and its principal recommendation is the ease with which it may be fitted. It is made as a right or left-handed lock, or with keyholes cut both ways, so that it may be used as either.

Wardrobe locks are specially made for the pieces of furniture they are named after, though in general principle they are closely allied to others. The chief difference is that they have in addition a spring bolt which is turned by a spindle connected with the fancy brass handle which is generally used on wardrobe doors. It may be interesting to note that till these brass handles came into vogue within recent years the spring latch was above the lock instead of as now below it. The principal wardrobe locks are the cut latch, the link plate, and the single bolt latch lock. This latter has only one bolt, which, when free, acts as a spring catch, and is secured by means of the key. The one bolt, therefore, serves both as a spring catch and lock bolt. These, as other door locks, are made right and left handed.

Nettlefold's piano lock is an ingenious contrivance for withdrawing the bolt as in a drawer lock, and at the same time allowing it to fasten a lifting top, as in the case of a box or piano fall. This is managed by two side catches shooting out from the main bolt when this is projected.

Spring catches are made in great variety for various purposes, but as they are simple and easily obtainable it is unnecessary to specify more than a few of them. The most important is probably the pedestal catch, which is similar to those of wardrobes, except that it has a spring merely and no key, the latch being withdrawn by means of a spindled handle. The 'cut' form is now generally used—the link-plate being an older fashion.

Bale's ball catches consist, as their name indicates, of a ball instead of an ordinary latch. From this peculiarity they may be used where it is only necessary to keep a door closed, so that it can be opened without turning a handle, as a slight pull forces the rounded bolt inside the casing. Flush rings and catches are very useful occasionally, as the ring or handle when not wanted to withdraw the latch lies flush or level with the surface of the wood to which it is fastened.

Flush bolts are used to fasten one door of a pair, in order that the lock fastened in the other one may act securely. These bolts are let into the edge of the stiles at top and bottom, or in the case of small doors at one of them only. The bolt shoots into a hole for it in the top or bottom of the carcase. Various contrivances have been devised to do away with the awkwardness of flush bolts, but none of them have come into general use or superseded the ordinary kind, though in some of them closing the second door automatically bolts the first one.

Castors are made in great variety, both as regards size, shape, and the material of which the bowl or wheel is composed. Castors with brass bowls are sometimes used, but those most generally seen are with vitrified china bowls, which are made either white, black, or brown, to harmonise to some extent with the colour of the articles on which they are used.

Socket castors are those in which the leg fits into a kind of cup, whence their name. The size of castor is reckoned by the diameter of the top of the socket. For these castors the wood should be somewhat cut away to fit within the socket, which should not be simply put on a gradually tapering foot. The amount to which the wood is cut away of course depends principally on the thickness of the metal. Sockets are to be had in various shapes, the most usual being round and square.

Screw castors have a screw and plate instead of a socket. The screw is driven into the leg till the plate fits against the bottom. In the plate are holes for further fastening it with ordinary screw-nails.

Pin castors are the same as the last, except that a plain pin takes the place of the screw. This gives them additional strength, and when they can be obtained they are generally to be preferred. The pin should fit moderately tightly within the leg, and the plate be well secured with screw-nails. The size of these and screw castors is reckoned across the plate, and, like sockets, they are to be had both round and square.

In connexion with these castors rims of brass are commonly used. They serve not only to diminish the risk of the leg splitting, but give a finish which without them is wanting. They are made in a considerable number of patterns, and in sizes to match the castors. They should be tightly fitted just at the ends of the legs and slightly sunk in them.

Dining-table castors are more direct bearing than the ordinary small ones, but beyond this and size there is little difference, as they are made both with sockets and otherwise. Very similar to them are pivot-plate castors, which may almost be considered the same thing, only smaller in size. They are sometimes preferable to the ordinary kind for such things as washstands, which are heavy in proportion to the substance of their legs.

Iron plate castors are only used where they are not seen, as in pedestal writing-tables, &c., where they are hidden by the plinth.

As there are obvious defects in the general shape of the ordinary castor, or, in case this is considered too sweeping an assertion, let me rather say, that their strength is not so great as it would be were the weight immediately above the bowl, many attempts have been made to overcome the difficulty by direct bearing. In castors of the direct-bearing class a ball working in a socket takes the place of the ordinary bowl. As a rule, however, these ball castors are very unsatisfactory in use, for they get clogged up with dust and fluff from the carpets till they are little more effectual than a rounded-end leg. Many of them, no doubt, are very ingenious, and I by no means wish to indiscriminately condemn all of them. I must, however, caution the amateur especially not to place too much dependence on the practical utility of most of them till they have stood the test of use. A mere trial in the hands to see that the ball works easily is not sufficient. The only ball castor which I can recommend is Wright's patent, which is superior to any other of the kind, and free from the objections which are commonly ·urged against them, and is doubtless as good a one as can be devised. As there is a good deal of 'old-fashioned' furniture now made, it may be well to say that castors with sloping sockets are sometimes useful. These are known as paw castors, presumably from the fact that they are, or rather were when they originated, often ornamented by giving them a resemblance to the paw of a quadruped.

Hinges are at least as important an item in cabinet brass-work as any of the foregoing, and some space must be devoted to a few of the chief varieties.

Butt hinges are those for which there is most employment, as they are constantly met with, and they are probably more used than all the others put together. When it is said that they are the hinges ordinarily used for doors, even those who do not know them under the name of butts, will recognise them as the common plain hinge with long narrow plates, which can be fixed to edges of wood. They are, of course, made in a number of sizes, and for furniture purposes are almost invariably of brass. For superior work they are to be had with the fronts, *i.e.*, the sides of the plates which are visible when a door is open, and the backs of the knuckles polished and lacquered. These, however, are not often seen, unless on things of the finest quality, as an ordinary well-finished brass butt is usually considered sufficient. The wire or pin connecting the two plates, as is well known, generally stops short at the ends, but in an ornamental kind known as knob or tipped butts, they are finished with a small brass knob or tip which relieves the crudeness of the plain hinge.

When fitting these, and indeed all other hinges, some care is necessary that the work should be done properly, otherwise the door or lid cannot hang properly or open and shut without unduly straining them. It seems almost unnecessary to say that one hinge is seldom used alone, at least two being required in almost every door, &c. This being so, it is essential that the pins should be in the same straight line. Were they to slope in different directions, or one to project further than the other, the slightest consideration will show that the door could not move easily and freely as it ought to. The novice, now knowing what is requisite, will have little difficulty in working to the following hints. It may be supposed that a door has to be hung; for those who can manage this can easily fit the hinge under different circumstances. A very slight acquaintance with furniture will show that a door may either cover the edges of the carcase ends or be within them, and, according to the arrangement, the hinges must either be sunk in the edges of the door or of the ends. The custom is not invariable, but, as a rule, when a door is within the ends, the hinges are sunk in it. When the door is outside the carcase, the hinges instead of being let into the back of it are let in the edges of the carcase ends. It may help the novice to remember which is the best plan, by reminding him that the hinges should be sunk in the edges and not on the flat surfaces. If a well-fitting door is wanted, the hinges should be sunk so far that they are, when closed, just flush with the wood, in other words, that the door can lie close even when the hinge is in place. The space cut away near the knuckle should be exact, but it is customary to cut away slightly deeper towards the other, or the inner edge of the plates when the door is closed, in order to allow for any slight projection of the screw-heads.

To mark out the width for the space to be cut, set the gauge to half the width of the open hinge, and scribe from the front of the door. By this means the hinge pin will be where it should be in ordinary circumstances, viz., just outside the surface of the frame.

Usually it will be found easier to fit the hinge to the door first, instead of the reverse; but, of course, if the latter plan should ever be the more convenient there is no reason why it should not be adopted.

When the hinges are fastened to the door, hold this as nearly as possible in its place and partly open, the free hinge plate being also open. Now just bore one nail hole, using one of those in the hinge as a guide, and then screw up. This will to some extent serve as a support while the bottom hinge is being treated in the same way. If the door hangs all right, and opens and shuts easily, the other screws can be driven in, but in case it does not, try again with one screw in each hinge. By making only one hole till it has been ascertained that the door hangs correctly the wood is left intact for the others in case any alteration is necessary. When arranging the position of the door it is a good plan to put a piece of glass paper, knife-cut veneer, or something equally thin under it. This prevents it fitting too tightly against the bottom, and may save some trouble afterwards.

Whenever appearance is of consequence brass screws should be used to fasten hinges on with. Attention to these little details wonderfully enhances the appearance of good furniture, and a careful cabinet-maker pays as much regard to them as to the more important structural work.

Back flap or table hinges are squarer in shape than butts, and are used to connect wide surfaces such as the flaps of tables. To the cabinet-maker they are next in importance to butts. As they are often not visible when fixed they are largely made of iron as well as of brass.

Card-table hinges are of two kinds. In one the hinge is let into the top of the table, and in the other, the older form, into the edges. They are so constructed that they lie flush with the wood in which they are sunk, the two plates being connected by a small one instead of with a pin. These hinges are used for counter tops and all similar purposes where the knuckle of the back flap would be objectionable.

Desk or bagatelle hinges may be compared to back flaps, as they open in the same way, but have long plates and short pins.

Screen hinges are made with double action, that is, they allow the folds of a screen to be moved in either direction. On account of their cost they are seldom used except on folding screens of the highest class.

Centre hinges were more used formerly than now. They are let into the top and bottom edges of doors, one plate being sunk in the corresponding part of the carcase.

Piano hinges may be regarded as very long butts, one of the plates being continuous, and the other in short lengths. They are used by cabinet-makers for a variety of purposes, such as fixing davenport lids, or anywhere, except doors, when it is considered that one long hinge will be an improvement on two or more short butts. They are necessarily somewhat expensive, so that they are rarely seen on common furniture.

Hinge plates of a decorative character are sometimes used, though they are not much in favour. They may either be part of the hinge, in which case the ornamental plate is fastened outside, or, as is more usual, they are separate, and are screwed on outside close to the knuckle of an ordinary butt. They sometimes look well, but the way in which they are employed is rather too artificial to be altogether in harmony with the spirit of the best modern furniture.

Escutcheon plates for fitting over keyholes, or rather on the wood surrounding keyholes, are frequently used in place of the ordinary thread escutcheon, which is merely a rim of brass shaped like a keyhole and forced into the wood.

Brass handles are made in an almost endless variety of patterns, and there is no doubt that by a judicious selection the appearance of almost any article of furniture is considerably improved by using them in place of wooden or other knobs. They should be fastened on with brass screws, most of them being prepared only for those with rounded heads. Being mostly made in Birmingham, the professional cabinet-maker will find it considerably to his advantage to be in communication with a cabinet brass-founder or factor there, as it is rare to find a retail dealer, except, perhaps, in the neighbourhoods largely frequented by London cabinet-makers, who keeps much of a selection. In connexion with these remarks, it may be said, common, low-priced brass-work, whether in the form of hinges, locks, castors, or the thousand and one odds and ends of handles, hooks, &c., is never advisable. Good brass-work frequently redeems an otherwise poor piece of furniture, while the appearance of many a well-made article is, in the eyes of those competent to judge, utterly spoilt by the use of common light brass-work. A satisfactory handle for a large sideboard connot be got for a few pence, nor can the cheapest castors be expected to bear much weight. Using inferior brass-work on good furniture is very much akin to 'spoiling the ship for the sake of a penn'orth of tar.'

It has, of course, been impossible to do more than touch on some of the more commonly used brass-work, and the novice must by no means consider that the list has been exhausted. Enough, however, has probably been said to enable those who wish to do so to extend their inquiries, and they will find that there is hardly an article of metal work required in furniture which is not readily procurable from a Birmingham cabinet-factor.

I am aware that in obtaining things from this source the amateur must necessarily be at a slight disadvantage when compared with the professional cabinet-maker, as the founders and factors as a rule do not care for retail orders, their business being confined to supplying dealers and cabinet-makers. In large towns there are generally shops where cabinet brass-work is sold, and through these the amateur will have little difficulty in obtaining what he requires, though he must be prepared to find many dealers who are not so competent as the special factors to give advice as to what will be the best to use in unusual circumstances.

CHAPTER XVII.

CONSTRUCTION — TABLES.

General Advice—Simple Fancy Table—Small Table with Shelf Below—Octagon Table with Spindled Rails—Square Tapered Legs—Small Round Table—Common Kitchen Table—Leg Writing-table—Table with Flaps — Brackets for supporting Flaps — Sutherland Table—Double Sutherland Table—Rule Joint—Card Tables—Dining-tables.

MOST of the principal elementary joints, &c., of cabinet work having been described, the novice who can make them and has a good comprehension of what is required should be competent to make almost any article of furniture in a workmanlike manner. I do not say that he will be able to do so without thinking or in a mechanical manner, for every fresh article which he may be called on to make will require the exercise of common sense and thought in its construction. Provided, however, that he will use his brains, the man possessed with any ordinary degree of mechanical ability will meet with no insurmountable difficulty. Of course, it is not to be supposed that the novice will attempt to make any really large or complicated piece of furniture, but even with these he must remember that they are little more than a number of small pieces or parts. If these each in turn are well made and accurately fitted, the workmanship of the whole can hardly be anything but satisfactory. I do not, however, recommend the novice to be too ambitious at first. Let him make simple things till he can work with neatness and exactitude, for if any of these are comparative failures the loss will not be serious; indeed, the loss of wood and time will be more than counterbalanced by the experience gained.

For these reasons the construction of a few typical articles of furniture will be described in order that the instructions which have been given in preceding pages may be put into practice. It is of course unnecessary to describe again in detail each operation, so that the following pages, it is understood, will not be of much use to those who have not previously learned the more elementary portion of the work. It may also be well to explain that though all the designs, it is hoped, will make up into such articles as will please the majority of readers, ornamental details are left to be added according to the discretion of the worker. The articles as shown may be principally classed among the good ordinary plain furniture of everyday use and sale. They will, therefore, no doubt from this last reason, be more useful to the professional cabinet-maker than things of a more elaborate character would be. It is not professed that all the designs are 'new and original,' as the object is rather to show how to make those articles of common use than to produce a series of new patterns. At the same time the designs are not taken at haphazard, but are carefully selected so that neither the amateur who makes for his own use, nor the professional who makes for sale, need fear that the lines are *outré* in character. To enumerate each article or variety of article which is made is of course impossible, as to do so would require many volumes the size of the present one. Those who do not find what they want, or cannot unaided cope with difficulties of construction, may be referred to the publishers' notice which appears immediately after the preface, and it is hoped by the means there suggested that every novice and amateur may have it in his power to make well-designed furniture of any kind.

Tables may be named first, not only because they are found in such an immense variety, but because in the simpler forms they are easy to make, and will afford many a useful lesson to the amateur and novice generally.

Fig. 167 shows a small fancy table, such as is commonly seen in drawing-rooms. Without wishing arbitrarily to fix on sizes and substance of wood for a table of this kind, it may be said that ¾ in. stuff will do for the top and 1½ in. for the legs. The framing may very well be of the same substance as the top, and if economy

Fig. 167.—Small Table.

of material is desired pine veneered will do. In such small articles it is, however, almost unnecessary ever to use anything but solid. The top may suitably be about 20 ins. long by 12 ins. wide and 2 ft. high. The shaping of the rails is purely a decorative detail, and does not affect the construction. This is of the simplest character. The rails are merely fastened to the legs by dowel or mortise and tenon, as may be considered the most convenient, and the top fixed down. In such a small thing as this probably it will be sufficient

CONSTRUCTION—TABLES.

to fasten it with glued blocks placed in the angles formed by the frame and the top. If the wood is quite dry, as the top is so narrow the shrinkage will not amount to much. It must, however, be understood that when a top is rigidly fixed there is a certain amount of risk of its splitting through shrinkage, and a better method will be to fasten it with screws. These, if the rails are narrow and thick enough, may be inserted from below, but with wide rails it will be more convenient to drive them in slanting from the inside, recesses being cut for the heads, as shown in Fig. 168.

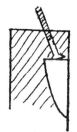

Fig. 168.—Screw through Rail.

Fig. 169 shows a table for similar purposes, but with the addition of a shelf below. This has a small ledge

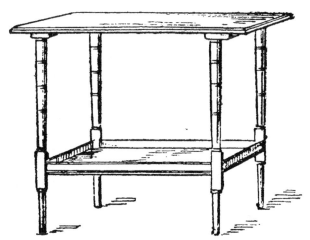

Fig. 169.—Small Table with Tray below.

of, say, 1 in. wide on its upper surface. The top, it will be noticed, is not supported by a framing as before, but, instead, the legs are fastened into pieces about 3 ins. wide, on which the top rests. The way in which the legs are secured will be seen in Fig. 170, which shows that they are turned with a shoulder and inserted in holes in the rails. If they fit tightly, glue alone will do to secure them, but otherwise a small wedge may be driven in to the top of each leg. This should be inserted in the contrary direction to the grain of the rails, especially if these are of thin stuff. The wedges themselves should be fastened with glue. This, however, must not be used to connect the top to the rails, which of course will be across the grain of the top. Were this to be glued down there would be great likelihood of the top casting; it would shrink, and, being bound by the transverse pieces, become hollow. The proper way, therefore, is to fasten it with screws, the holes for which in the rails should be sufficiently large to allow of some degree of play. I do not know that the amateur could derive a more useful lesson, showing the necessity for allowing for contraction or play of wood, than from this small table, or one of similar construction. It will do no harm to try the experiment, and this is more than can be said for most instances where construction has to provide for shrinkage. Let a piece of ordinary wood, not specially dried, be used for the purpose—fairly dry, but without extra precautions having been taken to ensure its being thoroughly so. The rails should be— for the sake of the experiment—just equal in length to the width of the top. Leave plenty of space in the screw-holes, especially those towards the edges, for presuming three are used, that in the centre may as well be a fixture. If the table is made under such conditions, and left for a time in a warm dry room, it will be found ere long that the top is shrinking and that the ends of

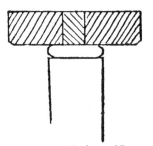

Fig. 170.—Fitting of Legs to Rails.

the rails project beyond it. These of course can easily be cut off, so that no harm will result from what to the amateur may prove a useful practical lesson. The wider the top the greater the amount of shrinkage, and the warmer and drier the place it is in the quicker the result will be.

The ends of the rails above the shelf are merely sunk in mortices cut in the legs. It will be unnecessary to make tenons, as if the ends are housed or let in to the depth of ⅛ in. to ¼ in. it will be quite sufficient. The corners of the shelf must be cut out to fit closely to the legs; it is then put in position and fastened by screws from below.

Fig. 171.—Table with Spindled Rails.

Fig. 172.—Fitting of Spindles.

Instead of plain rails a row of short spindles may be employed as suggested in Fig. 171, which represents a somewhat more ornate table. The spindles are shouldered and sunk at the bottom into holes prepared for them, and are capped at the upper end with a rail, as in Fig. 172. The rails are often placed below the shelf, in which case it is fitted as if it were a top.

As square legs are now much used, it may be well to note that when they do not taper directly to the end, but widen out there, as shown in Fig. 173, to a kind of foot, the easiest way

s

to form them is to taper them down first and then glue on the pieces to form the foot, as in Fig. 174. This method will be found much more speedy and easy than shaping the feet out of the solid.

Many small light tables are made with the legs sloping outwards, in order to give them more stability

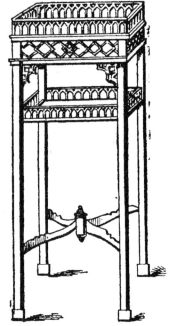

Fig. 173.—Square-legged Table.

Fig. 174.—Foot of Square Leg.

Fig. 175.—Small Table.

than they would otherwise have. Such a small table is shown in Fig. 175, where it will be noticed that the bottom board is fixed in a somewhat different manner, and fancy rails have been inserted between the end legs, the general construction being much as before.

In small round tables, as Fig. 176, the legs, which slope outwards, are best fixed into a round piece of wood smaller than the top. This is then screwed to it, though if care be taken to have the grain of both pieces coincident, there is little risk in using glue. The legs are fastened in the holes, which of course must be bored in a slanting direction, as already directed.

Larger plain tables are made in a similar manner, the only difference being that they are of more substantial material. Of these a common kitchen table shown in Fig. 177 may be taken as a typical example. In the illustration it is shown with square legs, and it may be noted that they are tapered from the inside only. To plane them off on each side would give them an awkward appearance. For any ordinary sized table of this description, say one from 3 ft. to 6 ft. long, the

Fig. 176.—Round Table.

Fig. 177.—Square-legged Common Table.

top and framing should be of at least 1 in. stuff, and a suitable width for the framing is about 5 ins. It will not do to fasten the top on with glued blocks, as it is almost sure to split if this course be adopted. Screws driven from the inside, as already suggested, are the best means of fixing, though it may be within the knowledge of some readers that nails driven through the top are sometimes considered satisfactory for pine work. The framing should be either tenoned or dowelled into the legs; and those who make use of their eyes will know that it is customary for the frame to be set back a little, say from $\frac{1}{8}$ to $\frac{1}{4}$ in., according to the size of the table, instead of flush with the surface of the legs.

Although a kitchen table has been named, it will be understood that any one of considerable size is made in very much the same way, the chief distinction between it and superior things of similar character being that these are made of choicer wood, and are more carefully finished. Thus, if instead of pine we take mahogany, line up the edges, run a suitable moulding on them, use turned legs, and put castors on them, we have a table fit for dining-room purposes, or any other use to which it can be put. Of course, as is well known, a dining-table is generally made to extend, but beyond convenience there is no reason why they should be so. Of them, however, more anon.

Now it will readily be understood that a table such as that last described can easily be made specially for writing purposes, though some modification will have to take place before the crude four-legged table is what is ordinarily understood in the cabinet trade as a leg writing-table. Drawers are usually placed in the framing, and the top, with the exception of a margin of from 2 ins. to 4 ins., is lined with leather, cloth, or some similar material. The number and position of the drawers depends entirely on the size of the table and the desire of the user. Thus it may have a drawer at one or both ends, or one or more on each side. A very useful form often seen has two drawers on one side, and a few hints as to the arrangements connected with these will be sufficient to enable the learner to put any number of drawers in any desired position. The front part of the framing instead of being solid is built up. The bottom portion forms the drawer-bearer, while the top rail or bearer is dovetailed into the legs and serves to screw the top to. The division separating the drawers is tenoned into the bearers. Extending to the back are the drawer-runners, and on them are the guides, all of which have been sufficiently described in the chapter treating of drawer fittings. The bearers may be from 3 ins. to 4 ins. wide, in fact anything that is convenient, and to economise wood are of pine faced up.

The table-top requires more 'making' than if it were plain, and it may be as well to say here that the construction of tops for plain pedestal tables is exactly the same. As no object would be gained by using mahogany or other superior wood for the part that is to be covered, or as it is generally called lined, it is made principally of pine. The edges have a piece of mahogany or whatever the wood being used is, of the same thickness jointed on, either plain glued or dowelled as described in the chapter on jointing up. Similar pieces are put on the ends, and those who have read the chapter just referred to will know that neither plain glued edges nor yet dowels will be the best method of jointing, as both the pieces are end grain. It may be noted that these ends should be put on before the sides, as after they are on, the edges can be shot truly to receive these. At this stage there is a pine top with a mahogany (or other) framing clamped on. The width of these pieces is unimportant. Some cabinet-makers make a practice of having them of precisely the same width as the veneer which forms the margin round the leather lining, and though there is little to be said against this it may be stated that such exactitude is not necessary when the work is properly done. Indeed, it may be urged, and not without some degree of reason, that it is better for the veneer to cover the joint on account of the extra strength gained thereby. It will not, however, do for the worker to trust to this too implicitly or think that because covered the joint may

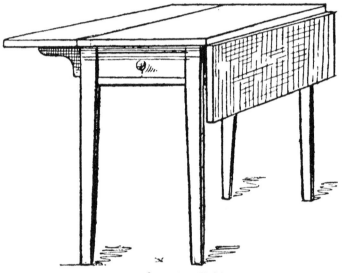

Fig. 178.—Flap Table.

be made in a careless manner. If it is not substantial and yields the defects will be visible through the veneer. When the joints are set this is put on afterwards, the grain throughout running in the same direction as the wood on which it is laid. It will be found convenient not to work the mouldings on the edge till the last thing, *i.e.*, after the veneering has been done. Lining the table tops is upholsterer's work, and does not concern the cabinet-maker.

Flap tables are very useful when it is desirable to economise space. They are known in the trade as

CONSTRUCTION—TABLES.

Pembrokes, Sutherlands, and by other names, which, however, do not concern the reader, as with few exceptions the lines of demarkation are not distinctly drawn. The great advantage possessed by flap tables is the ease with which the size can be altered, but with this must be coupled the slight disadvantage that they are seldom so rigid when open as an ordinary four-legged table.

In the commoner forms of flap tables, one of which is shown in Fig. 178, the edges are shot square and hinged by means of back flaps let in below. These must be sunk flush in most instances in order to allow for the bracket or support of the flap when up passing freely under them.

Fig. 179.—Folding Bracket.

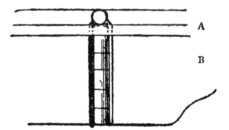

Fig. 180.—Knuckle Joint.
A, plan. B, elevation.

To support the flaps movable brackets are used. These when folded lie under the fixed top and against the framing, as shown in Fig. 179. The best form of hinging is that by means of the knuckle joint, which is shown in Fig. 180. It is, however, not altogether an easy one, though with care and by working to the following directions there ought to be no great difficulty in making it:—The piece of wood used should be preferably hard, nothing being better for the purpose than a good clean piece of mahogany. It must first be cut straight across, the sizes of each piece depending on circumstances. Now, on the edges of the ends which are to come together mark with the compasses a circle as nearly as possible the thickness of the wood in diameter,

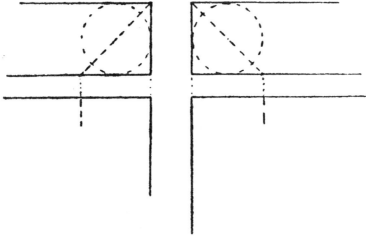

Fig. 181.—Setting out of Rule Joint.

as shown in Fig. 181. Then from the back corners mark off with bevel through the centres, as indicated by the

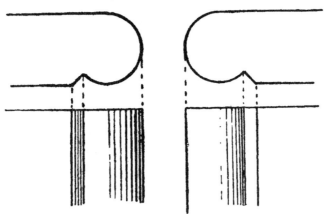

Fig. 182.—Shaping of Wood for Rule Joint.

dotted lines, gauge down the front in continuation of these and remove the wood down to the circle, giving as result the shape shown in Fig, 182. Next set off exactly the intersecting parts, and be careful when sawing these out to allow for the thickness of the saw in order that the hinge when brought together may fit closely. Some care will be needed when doing this to get the parts to work easily and closely without binding. When they are ready it merely remains to connect them with a stout wire pin or piece of iron rod, the hole for which must be bored accurately through the centre. If it is not so placed and perfectly straight it is useless to expect the joint to work properly. The shaping of the brackets is entirely a matter of fancy. The fixed piece is fastened on to the framing of the table by screws or otherwise. It may be well to caution the novice to see that the joint is fixed quite perpendicularly, as otherwise the top edge of the loose portion will incline upwards or downwards when opened outwards, with a corresponding slope of the table flap. In order to facilitate the drawing out of the bracket it is advisable to cut away a portion of the thickness, or at any rate to round the edge at some portion of the curve. If this is done the tips of the fingers can be inserted behind the bracket when it is folded against the frame and cause it to open more freely than it otherwise would.

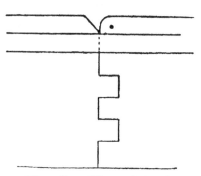

Fig. 183.—Finger Joint for Brackets.

A modification, and one that is more commonly adopted, as it is simpler and more easily constructed, is that shown on Fig. 183. In it the sockets on the loose piece are bevelled off, to allow the projections of the other to pass. This joint, it may be mentioned, is equally as strong as the other; its only objection, and that not a serious one, as the brackets are seldom visible, being that it is not so sightly.

Another way of fixing the bracket is by means of an ordinary hinge, though beyond its obvious simplicity it has not much to recommend it, and is seldom or never employed with good work. Such brackets, it may be suggested, do for folding flaps, which are often fixed to walls, equally as well as for portable tables.

When the width of the fixed or centre part of a flap table is great enough to allow of them being used, movable rails hinged with an iron pin, as shown in Fig. 184, afford an easy way of making a support to the

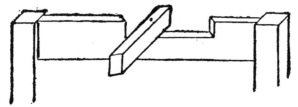

Fig. 184.—Pinned Rail for supporting Flap.

flaps. When these are folded or down the rails lie within the table framing, and are quite out of the way. It will be noticed that the ends are cut at an angle to form stops. The pin should be run through the rail and fit into the top as well as into the framing.

So far only short brackets have been alluded to, but a very small amount of consideration will show that they can be extended, and that a leg can be fixed to the free end. In this case the bracket becomes a rail; and, of course, the extra leg becomes an additional support to the table when the flap is up. Hence, flap tables so made are as rigid as they can be, and there is less danger of them being tilted up by leaning on the edge of one of the flaps. Fig. 185 represents a popular form of table of this construction, known as the Sutherland.

Fig. 186 shows a double Sutherland. It is a precisely similar table to that last described, but has an upper portion smaller than the bottom, and swing legs supporting the lower flaps only. Small brackets support those on top.

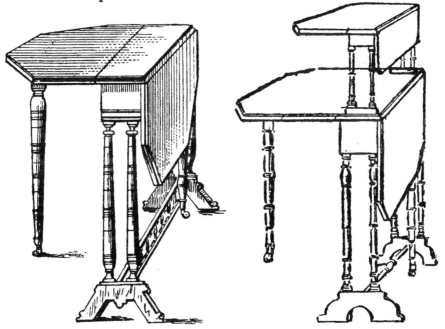

Fig. 185.—Sutherland Table.   Fig. 186.—Double Sutherland.

Those who are at all conversant with old furniture will recognise the resemblance to a method of construction which at one time seems to have been very largely adopted, and as it has many features to recommend it, an illustration is given—Fig. 187. It will be seen that there is an under framing into which the swinging leg is halved, the other short one being pivoted between the upper and lower rails, and a short footpiece fastened underneath. Sutherland tables are often made in small sizes for drawing-room purposes. The centre-piece for a small one, say 26 ins. × 22 ins., when open is from 4 ins. to 6 ins. wide, and of course for the size named is 22 ins. long. The blocks into which the tops of the end legs are dowelled are connected by one rail in the centre. On each side of this are fastened the fixed portions carrying the movable rail and leg. The bottom ends of the legs are fixed in splayed blocks, which may be shaped

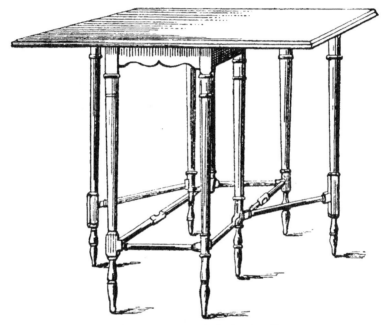

Fig. 187.—Table with Swinging Legs.

according to fancy. Castors may or may not be used below these, but in any case the swinging legs should have them. It may be noted that such small tables are usually called dwarf Sutherlands, and naturally many modifications of detail are found.

In these and all other flap tables of the better kind, instead of the edges being square where they are hinged, the rule joint is used. The former, though equally strong, is not a neat-looking form when compared with the other. The rule joint is shown on an enlarged scale in Fig. 188, the dotted lines representing the flap when down. It is by no means an easy one for unaccustomed hands to make; and however proficient the reader may be in general work, he certainly cannot be recommended to make his first attempt at forming a rule joint on a table. It will be far better for him to work it out first on waste wood, as unless the parts fit well and smoothly it is difficult, if not impossible, to make any improvement by tinkering at them afterwards. It will be noticed that the hinges are not visible even when the flap is down, the round edge of the fixed part not being at any time entirely uncovered by the hollow in the flap.

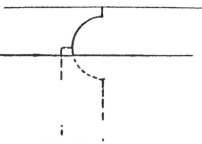

Fig. 188.—Rule Joint.

Unfortunately it is not an easy matter to explain the construction of this joint in such a way as to make it intelligible to those who are not acquainted with it, and the novice will do well if he can to get some friendly cabinet-maker to show him practically how to go about it, if he finds himself unable to follow the directions here given. However, if he will with tools and wood in hand work them out stage by stage, he will no doubt arrive at a very fair conception of what is wanted. It may be said that it is no uncommon thing to find a cabinet-maker who cannot make a rule joint properly, for it is seldom required except on such articles as are being described. Those who have much of this kind of jointing may provide themselves with a pair of table-planes, but the general cabinet-maker can do very well without them, using hollows and rounds and rabbet-plane instead. Table-hinges or back-flaps are the kind required, and care should be taken that the plates will work out to a right-angle with each other, in order that the table-flaps may hang properly. Some common hinges of the kind are not so accurate in this respect as they might be.

The ends and edges of the pieces to be connected being properly squared up, set the gauge to the centre of the pin, measuring from the back of the hinge or to a trifle more than the thickness of the plate. Now, on the end of the piece which is to be the centre portion of the table, scribe a short line with the gauge so set, and unless both pieces of wood are of precisely the same thickness gauge from the bottom. Next, with the compasses, describe part of a circle as shown in Fig. 189,

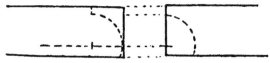

Fig. 189.—Setting out on Ends for Rule Joints.

the centre being on the gauged line. Then set the gauge to this centre from the edge of the wood, and scribe along top and bottom. The mark on top serves as a guide for rabbeting to, and the depth must be down to the circle. Lines for this of course should also be gauged, and it will be an assistance to set out the segment of the circle on both ends. The next proceeding will be to plane away the waste wood, but after what has been said in previous chapters no remarks about this can be necessary. The edge of the flap must naturally be worked to a hollow to correspond with the round.

When the edges are ready, the work of fitting the hinges can be proceeded with, and this will probably be found the most difficult portion of the work. They must be sunk into the wood at least flush to allow the swinging rail to pass, and the knuckles must be inwards, that is upwards, when the top is in position. Spaces for the hinge-plates must be cut, and a further deeper space for the knuckle. It will be noted that there is not much wood to spare when this is done. Beyond being careful not to cut through, so that the hinge is visible when the flap is down, the great point to be observed is that the hinge-pin must be on a line with the gauged mark on the bottom edge of the fixed portion of the top, and just the same depth in as the centre point from which the portion of the circle was marked with the compasses on the ends. To make this clear, let it be explained in other words that if a hole were bored from this centre point parallel with the edge and top or bottom of the table it should exactly meet the ends of the hinge-pins. By following these directions a moderate amount of practice should enable the learner to make a fair rule joint.

To prevent springing and to keep the parts rigid, the hinges should not be too far apart. Let three be used, say, for each flap of a table of the dimensions given. It will be well to fasten the hinges only temporarily till it has been ascertained that the parts work easily together.

To prevent the legs swinging further than is wanted, a stop should be fastened on the under side of each leaf.

In another variety of folding-table, but one not often seen, the top is fitted on a centre pivot which runs through a rail fixed for the purpose, and is fastened below by a nut. To support the flaps it is only necessary to turn the top partly round, so that they rest on the ends of the framing.

Many other varieties of folding-table might be named, but like the last they are so seldom seen that they cannot be regarded as ordinary articles of furniture. It may, however, be useful to suggest that in some the leaves instead of hanging are folded on top, and when opened out rest on slides. In another old variety the centre or main portion is free to rise and fall to a sufficient extent to allow the loose leaves to be pushed under it when not in use, and to sink to the same level as them when they are drawn out. In this case also the leaves are supported by slides.

The ordinary card-table is one of the most popular forms of construction in which the leaf, for there is generally only one of them, folds over on top. Such tables, as a rule, are about square when opened out, so that when closed the top is about twice as long as it is broad. The top revolves, *i.e.*, is fastened to the frame by a pivot on which it turns, so that when open the loose leaf is properly supported. The frame is usually covered in at the bottom, forming a space within which cards, counters, &c., may be kept. As the joint between the two parts of the top should be across the centre of the frame when the table is open, the novice would soon find that some method by which the position of the pivot can be easily and accurately determined is almost a necessity if tentative efforts are to be avoided. Perhaps the simplest means is that shown in Fig. 190, where the position of legs and framing is clearly indicated. The position of the top folded is represented by the heavy lines surrounding the frame, and open by the outer dotted lines, by which also the centre or joint between the two parts is seen. The other dotted lines show the way in which the quarter of the top is divided into squares to arrive at the point **X**, which is the position for the pivot. Below this a rail, also indicated in the illustration, is fastened across to the framing. Through it the pivot passes, and is secured by a nut underneath.

The tops of card-tables are almost invariably lined with cloth surrounded by a banding of veneer, in a similar manner to writing-table tops. As the loose leaves are usually veneered on one side and lined on the other, much care is necessary to prevent them casting. The ground wood should be as dry and even in the grain as possible. Not uncommonly, iron rods are sunk in the top in order, as far as possible, to prevent casting, but if sufficient precautions are taken other-

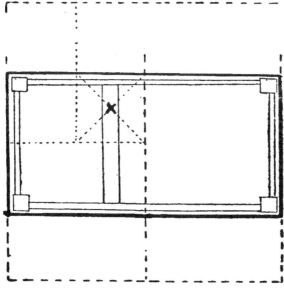

Fig. 190.—Folding Card-table Top.

wise it is doubtful if they are of sufficient advantage to compensate for the additional labour involved. When the edges and corners are square, the old-fashioned form of edge card-table hinges may be used, but with a moulded edge or round corners those which are made for letting into the surface alone are practicable. No special remarks about the way they are attached are necessary beyond suggesting that unless they are neatly let in flush with the tops it is almost as well to use hinges with knuckle showing, as small back-flaps, bagatelle-table, or desk hinges. Many other special forms of card-tables have been from time to time devised, but the one described is what is generally understood when a folding card-table is mentioned.

Dining-tables are as a rule made to extend, and are more massive than any of those which have been mentioned. When of large size and with telescope frames they are by no means easy to make, for all the parts must act easily, and before they can do this the most accurate workmanship is necessary. Probably a plain

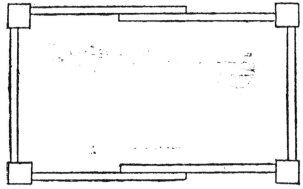

Fig. 191.—Simple extending Dining-table Framing.

tray-frame dining-table, which is also a telescopic one, inasmuch as it draws in and out in the same manner as the others, will be as much as the novice can manage. They have a simpler framing, and are equally as serviceable as the more complicated telescope arrangement, except for the largest sizes of tables. Perhaps it will be well to explain that what are known as tray-top tables in the workshop are generally spoken of outside as telescope tables, though strictly speaking they are, in a technical sense, not so. It is, therefore, quite possible that many readers may consider the construction here described as that of a telescope frame table.

CONSTRUCTION—TABLES.

As is no doubt well known, dining-tables are made to expand and contract by means of a screw worked by a large key. It is unnecessary to say more about these screws than that they can be got from any dealer in cabinet brass-work, &c., and that they are fitted to the end rails of the framing. The simplest form of extending dining-table may be gathered from the plan of framing illustrated in Fig. 191. From this it will be seen that there are two pairs of side rails, one pair fitting within the other, so that when the table is closed only the ordinary framing is seen. Such a table naturally does not admit of very great extension, and is chiefly valuable here as all tray-frame tables are made on the same principle. If it is wanted to draw out more extensively, three sets of slides or frames must be made, and so on. Within reason there is no limit to the number of slides, and, as stated, any but the largest dining-tables may be made in this way.

Now so far only the fact of the slides fitting against each other has been referred to, but any one can see there must be some connexion between them. The simplest form is to plough a groove, say $1\frac{1}{4}$ ins. wide by $\frac{3}{8}$ in. deep along one pair of slides, and fix a corresponding piece on the others, as shown in section Fig. 192. In this case a flat rail, say 3 in. wide by 1 in. thick, fixed to the under edges of each frame, will be necessary to keep them close. The rail on the inner framing may be so arranged that it forms a stop and prevents the two ends of the table being drawn apart. Of course the inner framing itself must clear the rail on the outer one. The arrangement may easily be made by dove-tailing the ends of the rail into the lower edges of the outer framing, making the inner frame so much narrower and fastening the cross rail on below its edges. The

Fig. 192.—Construction of Slides.

slides should be of good substance, say 1½ ins. thick and from 4 ins. to 5 ins. wide. The best way of connecting the framing with the legs is by dovetailing, as indicated in Fig. 88 (p. 158); but those who prefer may dowel or mortise and tenon.

The loose ends of the outer framing or slides should be sunk or housed to the depth of, say, ¼ in. in the legs when the table is closed. This will be neater in appearance than if they just fitted against them, and besides, additional strength is given through the frame being supported at each end by the legs instead of only at one. The strain on the slip which fits into the groove is relieved.

Instead of having a plain slide and groove, it is better to make these in dovetail form, as in section Fig. 193. This shape tends to bind the sliding frames to each other, but it must be confessed it is by no means easy for any but the most skilful to make a perfect fit. The parts must work together without sticking or binding on the one hand, and without any looseness to admit of irregular play on the other. With a short length this is comparatively easy, but the longer the slide the greater the difficulty. The groove is cut as before, and then dovetailed with a specially shaped rabbet plane, viz., one with a sole which is at an angle with the sides instead of square. The dovetailed piece requires no special tools, being merely planed on the bevel and then screwed on to the slide. With such construction the transverse pieces across the bottoms of the frames may be omitted, though no harm can ever result from using them.

For those who cannot make the dovetailed groove, it may be said that another plan is sometimes adopted which practically gives the same result in an easier manner. It is simply to bevel off three pieces of equal thickness, say ½ in. stuff, as shown in section, Fig. 193.

CONSTRUCTION—TABLES.

By fastening two of them to one surface of the frame, and the centre one to the part which works against it, it will at once be seen that a sliding dovetail is formed. This method may not meet with the approval of some cabinet-makers, but many good dining-tables have been made in which it is adopted. The slips should be glued and screwed on, the screw-heads being well sunk. It may be well to note that the grooves and dovetails should not be quite straight, but be raised a trifle towards the centre from the ends, *i.e.*, slightly curved upwards, so that when drawn out there is a slight tendency to lift the frame towards the middle. Not much rise is wanted, only sufficient to counteract what would otherwise be the natural drop. This of course is greater in a long than in a short table, so that it is impossible to say exactly what slope should be given to the grooves, &c.; but as some idea, $\frac{1}{8}$ in. may be named for a table drawing out to 8 ft. long. The easiest way to get a graduated rise and one equal in corresponding parts is to prepare a piece similar to a straight edge, and use it as a template.

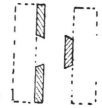

Fig. 193.—Dovetailed Slide.

Although there is no absolute necessity for them being so, custom decrees that the loose leaves of a dining-table should be 18 ins. to 20 ins. wide. Reference has already been made to the edges being dowelled, and it is only necessary to say here that the dowels and holes should be regularly placed so that the leaves are interchangeable, though as a matter of fact it is usual to mark the edge of each leaf, showing its natural position. To lessen the friction of the upper edges of the frame as they work under the fixed portions of the top, it is better to raise these slightly by pieces of veneer glued on to the top edges of those parts of the frame to which they are attached. The veneer gives sufficient clearance, and the slight rise is not observable. The top, it seems almost superfluous to say, must not be glued down, but be fastened with screws.

To prevent any of the slides being withdrawn various devices are adopted, and there is little to choose between them if the stop is satisfactory. One good and neat method is by means of brass or iron plates, say $\frac{1}{8}$ in. thick and measuring about $1\frac{1}{2}$ ins. long by 1 in. wide. These are fixed on the lower edges of two touching parts of the frame with a portion projecting, so that on the slides being drawn out as far as intended they come in contact with each other. Those who prefer some other form of stop will have no difficulty in devising one.

Ordinary and miscellaneous tables have now been sufficiently referred to. There are many other kinds, such as pedestal writing-tables and dressing-tables, which will be found mentioned elsewhere; and in concluding this chapter it may be said that few if any leg-tables are made in which the principles of construction vary to any considerable extent from those which have been named.

CHAPTER XVIII.

BEDROOM FURNITURE.

Plain Hanging Wardrobe—Small Wardrobe with Drawer—Wardrobe with Straight Ends — Beaconsfield Wardrobe — Six-ft. Wardrobe with Long Trays and Drawers—Short Trays and Drawers—Fittings—Toilet Table Glass—Dressing Chests—Washstands—Pedestal Cupboard—Pedestal Toilets.

THE furniture of the bedroom consists principally of wardrobes, dressing-tables, washstands, chests of drawers, and pedestal cupboards. Some may be inclined to think that the piece of furniture which gives its name to the room, viz., the bedstead should have been included in this list, and it may be well to explain that brass and iron bedsteads are now so much used that those of wood are practically obsolete. The construction of those of metal, of course, does not concern the cabinet-maker, so that, for present purposes, it is not necessary to consider them. The bedding, of course, so far as mattresses are concerned, comes under the upholsterer's hands, and, consequently, will not be described here. Following the course adopted in the last chapter, typical examples of the principal furniture will be given, details of design being left to the skill and discretion of the maker.

Chief among the things constructed by the cabinet-maker is the wardrobe, of which many varieties are found in size and arrangement. In most of them the leading features are a cupboard for hanging things in, with, when size permits, drawers, shelves, and sliding trays, and it is generally, invariably when ladies are concerned, considered that a large looking-glass is an essential feature. This, of course, is introduced as a door panel. As even a small wardrobe is a somewhat

280 BEDROOM FURNITURE.

cumbersome piece of furniture, which it would be difficult to move to or from a room, while a large one could

Fig. 194.—Small Hanging Wardrobe.

hardly be removed entire, it may be as well to say at the outset that wardrobes, like other big articles of furniture are built up of separate parts which are fastened up together.

The smallest wardrobe is the plain hanging variety shown in Fig. 194. It consists of three parts, the main or cupboard portion, plinth, and cornice, and requires few remarks. The width seldom exceeds 3 ft., height and depth from back to front being regulated according to ideas of convenience. The door is shown with a comparatively small glass, as to have one to the bottom would, on account of the weight, be rather a strain on such a small carcase, though, if this is made sufficiently heavy and deep from back to front, there is no reason why it could not carry a much heavier door. This, it will be noticed, does not extend the whole width of the front. The remainder is taken up with solid pieces, to which something of the appearance of framing and panelling is given by mouldings. Of course, these pieces may be actually framed and panelled, but the width is so trifling that it is hardly worth while doing so, while, on the other hand, the mouldings may be omitted, and the parts left plain, or be relieved by scratched beadings. The ends are connected by solid top and bottom lap-dovetailed into them. Plinth and cornice are secured by means of square blocks fastened to the top and bottom, and fitting into the corners. Thus, to fit up the parts, the plinth is laid on the ground, the carcase placed on top with the blocks fitting into the corners of the plinth, and, finally, the cornice is laid on. The back is munted and let into a rabbet in the ends, and lies over the edges of top and bottom in ordinary work, while if something better is wanted it may be panelled. Fig. 195 represents a hanging wardrobe with drawer at bottom, and is available for sizes up to 4 ft. or even 4 ft. 6 ins. wide. It is made in four parts, viz., the lower portion containing the drawer, the cupboard, and, of course, cornice and plinth, though this latter, if it is considered preferable, may be fastened to the drawer carcase. In the main the construction is the same as before. The lower portion is merely a case to contain the drawer, which usually

Fig. 195.—Hanging Wardrobe with Drawer.

does not work direct against the ends, the thickness of which in front is apparently increased by pilasters of say, 2 ins. to 3 ins. in width, the space behind them, of course, being vacant except for the guides which are

Fig. 196.—Wardrobe with Straight Ends.

necessary for the drawer. The upper carcase is made as before, but the fixed pieces of the front may as well be framed up and panelled. The top and bottom, it will be noticed, are covered by the door, which may be either hung with butts or with centre hinges. In the latter case either the door must be rounded at the back edges by the top and bottom, or a small space must be cut in these to allow the door to hang. If centre hinges are used, a beading strip may be fastened on to the fixed portion similar to the one on the edge of the door. These pieces being fastened with one edge projecting serve to cover the joint of the door. A wardrobe of somewhat different construction, inasmuch as it has three drawers in the lower portion and has straight sides, is shown in Fig. 196. Owing to the absence of projecting plinth and cornice, this form is admirably adapted for fitting into a recess, as the sides fit close against the walls, and no space is wasted. When this method is adopted, the cornice, of course with straight sides, may be made separately, but it is usual not to do so, the moulding in front being simply let in between the two ends. The top cannot be dovetailed unless it is above the cornice, an unusual form of construction but one which gives additional height within the cupboard, and must be fastened by tenon or other joint into the ends.

The plinth may also be made separately, but it will occur to any one that it will entail less work if formed like the cornice. Screws are used to fasten the two main portions of the wardrobes together.

Fig. 197 represents a popular form of wardrobe known as the 'Beaconsfield.' On the left-hand side it has a hanging cupboard enclosed by door with glass panel; on the other are drawers below and a small cupboard above. In this may be either sliding trays or fixed shelves. Immediately above the plinth is a long drawer. Such a wardrobe may be made in three or two

Fig. 197.—'Beaconsfield' Wardrobe.

carcases according to size. If made without the long drawer at the bottom, two will be sufficient.

The ordinary form of three-door wardrobe is shown in Fig. 198. Two-thirds of the inside space, that is, a

Fig. 198.—Six-ft. Wardrobe.

portion enclosed by two doors, is usually occupied with drawers below and sliding trays above, the remaining portion being a hanging compartment. As this form of wardrobe may almost be said to be the standard, a little more space may be devoted to it and its fittings than has been to the others. Of course any of the fittings, when circumstances allow, can be modified to suit to any of the wardrobes of smaller size or different shape.

The three carcases in which such a wardrobe is made are sufficiently shown in Fig. 199. The lower left-hand one is practically a plain chest of drawers without the ornamental adjuncts of plinth, finished top, &c., and if

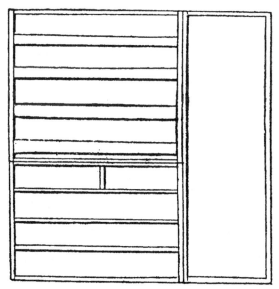

Fig. 199.—Interior of Three-door Wardrobe.

the doors are hinged to cover the ends the drawers run against these. As has been stated, in such constructions as this the top and bottom of the carcase should be a trifle longer at the back than in the front in order that the drawers may run easily. The carcase above the drawers contains the sliding trays. These are in reality nothing but a modified form of drawer, being comparatively shallow, and usually having the front much narrower than the back and sides. They may also be compared to sliding shelves, with a rim round them to prevent things falling off. Necessarily from their formation a different arrangement from that of drawers is necessary for the carcase arrangements for fitting them in, bearers, runners, and guides being dispensed with, or rather they take a different form. The trays may simply rest on ledges fixed to the carcase ends, but in this case there is the obvious disadvantage that there is nothing to prevent them tipping up when partly drawn out. This mishap may be obviated by ploughing a groove in each tray side, and fixing a slip of wood to correspond into the carcase ends, as shown in Fig. 192. The strips are best secured with screws, the heads of which should be well sunk. The groove in the tray sides may come right through to the front, but this being fastened with the lap-dovetail joint, it will look better to stop them short just behind the tray front. Additional strips may if desired be fastened on the ends for the bottom edges of the trays to run on, but as a rule they are not necessary.

Another method of securing the trays, but a somewhat clumsy one, is to fasten strips above as well as below them. Its chief merit is simplicity.

The fittings of the right hand or hanging compartment vary considerably, according to the ideas of the maker or of the user of the wardrobe. Hooks, of course, to hang things on are naturally essential, and may be looked upon as the principal fittings, to which the others are subordinate. Special wardrobe hooks are among the articles made and sold by the cabinet brass-founder. They are usually fastened, not direct on to the ends and back of the carcase, but to rails, which are then screwed on. These afford support for a shelf to be laid on them if one is considered desirable, and in this case they must be fixed sufficiently low down to allow the required space above.

Instead of the hooks being on fixed rails, they may be fastened on to the insides of a sliding frame. The advantage of this is that in a large compartment a fourth row of hooks may be used instead of only three, as there is now a front rail available. The frame is made like a drawer or sliding tray, but of course has no bottom.

Instead of, or in addition to, the ordinary hooks, a cluster of revolving hooks may be hung from the top, but these are not often used.

Another method, and often a very convenient one, as by it great economy of space is secured, is by brass rods from side to side, and having sliding hooks on them. These hooks are very much like those often used to hang pictures to rods with, but are lighter, especially on the hook portion below the ring.

Very often at the bottom of a hanging compartment is a deep covered-in receptacle, commonly known as a bonnet-box or drawer, according to its formation. The maker must use his discretion whether there is sufficient space left for hanging purposes if both shelf at top and drawer below are fitted. It may be observed that it is comparatively seldom that both are required.

When a drawer is made it is only necessary to say that it must be sufficiently deep, and that immediately above it and covering it is a shelf or top fitted into the carcase ends. From their size such drawers are somewhat unwieldy, and it is no uncommon thing to find them dispensed with in favour of a fall-down front. This is simply a door hinged to the bottom of the carcase, and fastening by a lock or spring bolt into the shelf above. To prevent its casting, such a front should be either framed and panelled or clamped at the ends, this latter being the more ordinary form.

Another method is to fasten the front in, and have a sliding tray above it to give access to the interior.

A variety of large wardrobe (6 ft.) precisely similar in external appearance to the one last described is somewhat differently fitted inside, though it also has drawers and trays. These, however, are short, the wardrobe being formed of three upright carcases, two of which, generally those at the ends, are hanging, and the other contains the drawers and trays.

Many fancy forms of wardrobes have been made, especially of late years, but the essential differences of construction are principally in design, however much they may vary in shape and general arrangement. It may be said that the insides of hanging compartments of wardrobes are generally lined with striped and glazed lining, so that the objectionable ochre colouring so often put on by London trade makers who sell principally 'in the white' or in an unfinished state should be omitted, as otherwise it must be washed off before lining. In superior wardrobes the covering of the glass behind is usually framed and panelled, but in the commoner kinds thin wood, either running from top to bottom, or if the size is considerable with a munting across, is generally deemed sufficient.

Toilet or dressing-tables and washstands are, so far as the lower parts are concerned, almost identical, the differences being more modifications of design than anything else. In general construction they are similar to ordinary leg-tables. Unless in very small or common cheap toilets—as both washstand and dressing-table are commonly spoken of when both are referred to—it is usual to connect the end pairs of legs, and fix a board or shelf to the rails. As a double washstand—though this is not a recognised designation among cabinet-makers—may be wanted by many readers, it may be well to say that 4 ft. is usually looked upon as what is generally considered as such, though of course larger sizes are frequently met with. An ordinary full-sized suite may be said to consist of 6-ft. wardrobe, 4-ft. toilet-table and 4-ft. washstand, pedestal cupboard, towel-rail

Fig. 200.—Toilet Table with Glass.

and three chairs. The size is taken across the front, and the width of a 4-ft. top may be about 2 ft. The usual height is about 2 ft. 6 ins.

The dressing-table is now almost invariably made with the glass attached, instead of, as it used to be, a separate article. As is no doubt well known, the glass frame is so hung on standards that it can be swung to any desired angle and fixed there. Fig. 200 represents a 4-ft. toilet-table of ordinary construction, with jewel drawers and glass attached. After what has been said about tables, no remarks about the lower part can be necessary. The jewel drawers and the cases containing them are of very simple construction, the ends being connected with lap dovetailed top and bottom of pine. Below the bottom is a lining with moulded edge, mitred at the corners, while the outer top is screwed on from the inside. The bearer between the two drawers may be of narrow stuff as usual, but as the size of these parts is so small it may as well extend as far back as necessary, so that side runners are not required. The standards supporting the glass are turned columns left square to a short distance above the jewel boxes, at the corners of which they are let in and must be firmly secured. The small bracket gives a finish, and forms a ledge to the top of the box.

Another form of support to the glass is shown in Fig. 201, where it will be seen that the turned column has been replaced by a shaped bracket, and the single drawer on each side by two. This may either be screwed on to the top of the box or be a continuation of the back.

It must not be understood that the drawers, &c., on a dressing-table back are always of the same arrangement, for there is considerable variety. To describe and illustrate even those which are most commonly met with is impossible here, but among them may be suggested—one drawer instead of two on each side; drawer or drawers raised above the top, leaving a space beneath; drawers with shelf above, small cupboards, &c.

The glass frame hardly requires remarks beyond saying that a simple and easy way of constructing it is to make a pine frame of sufficient width and thickness, and

Fig. 201.—Toilet Table, with Brackets supporting Glass.

face the front to form a rabbet within which the glass lies. A pine frame, which, of course, is not to be compared with one made solid, should have the outer edges veneered or faced, and can only be recommended for cheap work. The top moulding which is shown is merely an ornamental detail, and may be either fixed on the face of the top of the frame, which at that part must therefore be wider than elsewhere, or be screwed on the top edge. In either case the return on the ends should be mitred at the corners. Like the moulding, the pediment above it is merely ornamental, and not only in dressing-tables but in other furniture is useful as a means of finishing top edges less abruptly than when straight lines alone are used. Being above the eye it can be screwed on from the top edge.

The position of the glass movements by which the frame is connected with standards is of some importance. The best plan in ordinary circumstances is to fix them so that the frame is fairly balanced, with if anything a slight excess of weight below them.

Dressing-chests are merely chests of drawers usually about the same height as tables; and, therefore seldom containing more than two long and two short drawers, as in Fig. 202. For small toilets, especially where jewel drawers on each side could not be placed without either making them unreasonably little or unduly curtailing the size of the glass, it will be found very convenient to have one long drawer along the top and under the glass.

As it will not again be necessary to refer to chests of drawers, it may be said here that the plinth is almost invariably a fixed one. It should be said that, as the backs of toilet-tables are often placed near a window so as to be visible from the outside, they should be neatly finished, especially behind the glass.

Washstands usually have marble tops instead of wood, and are fitted with tile backs. The marble is got ready from the marble mason, and the cabinet-maker has nothing to do with it except give the sizes. The kinds generally used are the ordinary white with grey

BEDROOM FURNITURE. 295

veining and St. Anne's, which is just the reverse, being a dark grey or almost black with white or greyish veins. Other varieties are occasionally used, but not often; the principal one beyond those named being probably Sienna, the prevailing colour of which is reddish.

As the lining, which would ordinarily be fastened to

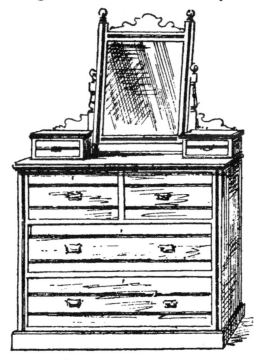

Fig. 202.—Dressing-chest.

the top, cannot well be attached to the marble, it is fixed to the table instead. Though it may be omitted it is not usual to do so, if the dressing-table has a lined top; and it may also be hinted that it is desirable, too, for mouldings on both toilets to be alike if the two articles are uniform in details of design; or, as the cabinet-maker would say, are a pair.

The ordinary tile back is of very simple construction,

whether with one row of tiles, as in Fig. 203, or with two or more. It consists of little more than a framing with an upright at each end. The framing may be either separate and fastened to these, or they may be rabbeted out behind, and form the ends of the frame.

Fig. 203.--Washstand, with Single Row of Tiles in Back.

In setting out tile backs a good deal depends on the size of the stands, as the tiles are only made in certain sizes, the 6 ins. square being the ordinary one. Seven of them are required for a row for a 4 ft. washstand, so that it will be seen the frame is not necessarily the whole length of the marble, as if this were insisted on the frame would often have to be made heavier than consistent with symmetry. In order to support the back to some extent bracket-arms, as shown, usually extend part way to the front. They are screwed through from the back of the framing. This, complete, is fastened on top of the marble by screws through holes prepared by the marble mason. The tiles themselves are usually just placed within the rabbet, and covered over with a thin wood backing. Their size being known, it is a simple

Fig. 204.—Washstand, with Pedestal Cupboard under.

matter of calculation to dispense with the blocking which is necessary generally for glass.

In the illustration (Fig. 204) a pedestal cupboard is shown under the stand. This is a combination now often met with. The cupboard is perfectly plain, *i.e.*, without either plinth or finished top, as both are unnecessary. The bottom is screwed to the board below, and the top is fastened to the drawer bearer or a runner as may be most convenient. In this washstand it will be noticed that a towel-rail is attached on each side. It consists of a piece of brass tubing fixed to brass arms which are sold for the purpose.

Fig. 205.—Pedestal Washstand.

When the pedestal is separate it is almost equally simple in construction, the principal difference being that it has a plinth fixed and a covering top, the inner one connecting the ends being of pine, and dovetailed to them as usual in carcases. One shelf supported on rails fixed to the ends is generally fitted inside.

Fig. 206.—Pedestal Toilet-table.

The construction of ordinary towel-rails is so simple

that nothing need be said of them. The chairs, of course, do not come within the work of the cabinet-maker, and those who wish to get them for trade purposes will know that they are mostly made in and about High Wycombe.

Pedestal toilets differ principally from those described by having pedestals instead of legs, as represented in Figs. 205 and 206. As will be noticed, the pedestals of the dressing-table have drawers, while those of the washstand have doors to form cupboards. This is the usual arrangement. In the backs or upper portions it will be seen that there is considerably more work than in the examples previously given. The dressing-glass is shown within a fixed frame, while the boxes at the sides are of more elaborate design, and the drawers are replaced by doors hinged at the bottom. The wash stand has a double row of tiles with shelf above, and in the centre a frame containing a looking-glass. It may be well to say that these upper portions are not peculiar to pedestal toilets, being given in connexion with them here by way of variety. The lower part of each must be made in one portion.

Pedestals may be made of the ordinary construction with plinth instead of feet, and the stands may be either in one or three portions, viz., the two pedestals with fixed plinths, and the top framing connecting and containing the three top drawers.

CHAPTER XIX.

LIBRARY AND OFFICE FURNITURE.

Pedestal Writing-table—Double ditto—Desk Slopes—Register Writing-tables—Cylinder Fall-tables—Old Bureau—Dwarf Bookcases—Secretary Bookcase—Nests of Pigeon-holes.

WRITING-TABLES or desks of various forms and bookcases are naturally the principal articles of library or office furniture.

In a writing-table the requirements are that it should have sufficient stability, and be provided with accommodation for papers, &c. The ordinary leg writing-table, with lined top and drawers in the frame, has already been sufficiently referred to. So far as convenience for writing on goes, nothing more could be wanted, but it is deficient in accommodation otherwise.

The pedestal writing-table, as shown in Fig. 207, is more useful for ordinary purposes. It is made in three parts, viz., the two pedestals, each containing three drawers of graduated depths, and the top containing three drawers. The plinths are fixed. The castors used are the iron-plate variety. Many very commonly made pedestal tables have the backs nailed or glued on, and are afterwards covered with knife-cut veneers. A better way is to frame and panel them. Instead of standing on plinths they may be made as suggested by Figs. 205 and 206 of pedestal washstand and toilet-table. A very ordinary size for these tables is 4 ft. by 2 ft. 3 ins. top, and 2 ft. 6 ins. high. Instead of drawers in both pedestals, one is often fitted as a cupboard with door. The top itself is made as for a leg writing-table, viz., principally of pine, with veneered banding. Double pedestal writing-tables are usually longer, and considerably wider in proportion to their length, in order to provide accommodation for two writers. In this case they are double-

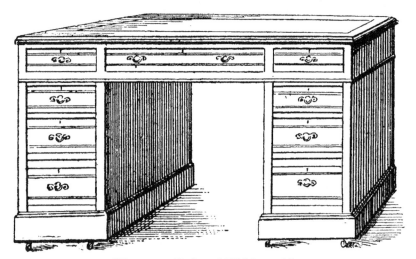

Fig. 207.—Pedestal Writing-table.

fronted, the usual arrangement being for each pedestal to have drawers opening to one front and a cupboard to the other, a partition separating the two. The cupboard, it made be said, is generally to the left hand of the

Fig. 208.—Desk Slope.

sitter. On top of each pedestal are four square blocks which fit accurately in the corners of the top framing, which is laid loosely, *i.e.*, without being fixed by nails or glue. Care must be taken not to have the blocks so

LIBRARY AND OFFICE FURNITURE.

high as to interfere with the free action of the drawers above them.

On flat-top writing-tables desk slopes are often used. They are simple in construction, as will be seen from Fig. 208, which represents a single one. When made double the flat top is widened, and another slope added on the other side. The slope may either be made to lift

Fig. 209.—Pedestal Register Writing-desk.

up so as to form a receptacle for papers underneath, or be fixed to the ends. In the latter case it need not have any bottom. Occasionally, but very rarely, they are fastened to the table top.

The register writing-table is shown in Fig. 209. As will be seen, it is very like the last named, with the addition of two small cupboards or nests of drawers on top, with a sloping writing surface in the middle above the space between the pedestals. The standard size may be taken as 5 ft., the width being in proportion. The

304   LIBRARY AND OFFICE FURNITURE.

pedestals are made as before. The top is usually made to show as three drawers in front, but the centre is a sham one, the space under the slope being a deep well to the bottom of the top, and often fitted behind with small drawers or pigeon-holes. The well is made separately and let in through a hole in the top, to which it is afterwards fixed. The slope is made like a small top with clamped ends, and veneered banding for lining. On the flat top, which otherwise is made as before, the banding of veneer goes round each portion. Each set of drawers on top is secured by one lock and key fitted to a hanging pilaster hinged on to the outer edge of the casing. The other is

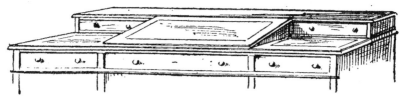

Fig. 210.—Upper Portion of Half Register.

fixed, necessitating comparatively wide runners. Behind the hanging pilasters the drawers go right to the ends, so that when it is locked, the lock plate being fixed to one of the drawer bearers, the drawers cannot be pulled out. To allow the pilaster to fit close to the fronts, the edges of the bearers behind it are cut down to their level. The rest of the construction speaks for itself.

A modified form for upper part of this writing-table is shown in Fig. 210, and is often known as the half register. Like the former it has a centre well, but the top fittings are of simpler character, as will be seen from the illustration.

Both these tables are sometimes made with legs instead of pedestals, and are then known as leg registers.

The cylinder-fall writing-table, of which the upper portion is shown in Fig. 211, has the advantage that by pushing back the writing or table part and pulling down the fall, papers, &c., can be instantly put away without disturbing them. On the other hand, there is the slight objection that it is a somewhat clumsy-looking and cumbersome piece of furniture. It is made in three parts, two pedestals and top. It is the construction of this latter which differs from that of other pedestal tables, and it may also be said that it is not one suitable for the novice, as the most accurate workmanship is essential. The front portions of the ends are the seg-

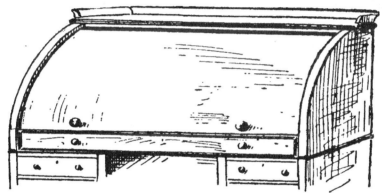

Fig. 211.—Top of Cylinder-fall Table.

ment of a circle, the top or flat part in continuation being at least of such width as to admit of the cylinder-fall being pushed sufficiently back. The exact sizes can easily be got by full-sized setting out. The fall may be troublesome to those who have never seen one made. It is formed of pieces of board, say 3 in. to 4 in. wide, with their edges bevelled and joined together, after which it is further rounded off. Unless the pieces are unusually wide, the inside may be left as it is. The easiest way to get the curve rightly adjusted and the pieces properly fixed together is to make a light framework somewhat similar to that used by builders when constructing an arch. It can easily be made of any rough

X

stuff, and, of course, the sweep on its edge must be exactly the same as that of the fall. With this frame as a guide, it is simply a matter of bevelling the edges of the boards to get the groundwork of the fall correct. After it has been smoothed and rounded it is veneered. The ends of the fall may work in grooves in the ends of the top, but a very much better plan is to hinge them. The hinges, if they can be called so, take many forms, and it will be better to say that the principle is as shown in Fig. 212 than to enumerate these, as the alterations are merely in detail. Almost any piece of hoop-iron may be used for the purpose, as the illustration will serve to show. The edge of the fall is there represented with the irons placed. These meet and are hinged by means of a screw, nail, or iron peg driven to the ends. Care must be taken that this pivot is exactly in the centre of the circle of which the fall forms part. The irons work behind thin inner ends, fixed, with a sufficient interval, inside the outer ends of the top. It must be noticed that the slightest irregularity in fixing the fall will cause it to work stiffly, if not to jamb entirely.

Fig. 212.—Method of hanging Fall.

The table part is usually framed up, and slides within the ends. In the ends of the table grooves are ploughed, and in them fits a corresponding piece of tonguing. The centre of the table is hinged in front to form a slope when necessary, and when up is supported by two short pieces of wood loosely screwed inside, so that when not wanted they fold out of the way. To prevent them slipping the loose ends fit into notches cut for them. In a similar manner the spaces on each side of the slope may be made with lids to lift up.

LIBRARY AND OFFICE FURNITURE.

The interior fittings, pigeon-holes, &c., are made separately inside a loose case, which is fitted in place when making the job up.

Instead of a solid cylinder, a shutter or tambour fall, built up of narrow pieces of wood, not glued together but stuck on a backing of canvas, so that they are flexible and will adapt themselves to any reasonable curve.

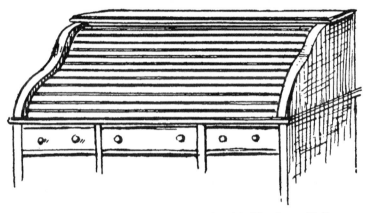

Fig. 213.—Top of Writing-desk with Tambour Fall.

Such falls run within grooves in the ends and must be very accurately fitted, sufficient space being left behind for the falls to be pushed up. The width of the pieces must be regulated according to circumstances, and by rounding the edges of each or working beads on them a handsome appearance may be given. Fig. 213 represents such a desk with the fall partly closed, and it will be seen that a different shaping has been given to the front. The table top also, it may be noted, is fixed, and does not pull forward. Most of the American writing-tables are made on this principle, and it must be admitted that many of them are far ahead in point of convenient arrangement of the interior to those generally made in this country. The interior fittings of any writing-table, it should be said, are entirely a matter of fancy on the part of the maker, and those who have an opportunity of doing so will do well to examine the American arrangements.

The bureau illustrated in Fig. 214, though an old-fashioned thing, is not without its advantages. Com-

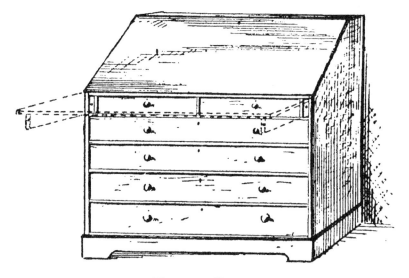

Fig. 214.—Bureau.

pared with other writing-tables, &c., it is not often made, but sufficiently so to justify mention here. The lid is hinged to form, when open as indicated by the dotted line, the writing space, and is supported by sliding rails at the ends. The usual arrangement of the lower part is drawers, but any that may be preferred can be adopted, the simplest naturally being a plain cupboard. The lid is clamped up at the ends or framed all round, the panel on the inside at any rate being flush. The top or front edge and ends are rabbeted to rest against the top and on the ends of the carcase, and the only point to which

LIBRARY AND OFFICE FURNITURE. 309

the novice's attention need be directed is the necessity of allowing for the thickness of the top, which lies within them, by not starting the slope of the ends directly from the top.

Bookcases are made in every conceivable size and shape, the simplest being that known as the dwarf bookcase, of which a small one is represented in Fig. 215.

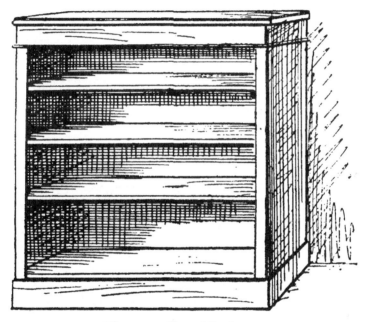

Fig. 215.—Dwarf Bookcase.

It consists of little more than an open carcase supporting the shelves. These, as is well known, are generally made movable and adaptable to any desired distance apart. The usual mode is by means of pieces of wood shaped and fixed to the ends, as shown in Fig. 216. Movable rails on which the shelves lie are fitted to these, the corners of the shelves being cut out to fit. When desired, dwarf bookcases can have doors fitted to them, and the edges of the shelves look best when finished with leather edging. This may be stuck direct on to the edges, or on to slips which are sometimes glued to the under surface, and sometimes sunk in grooves ploughed for them.

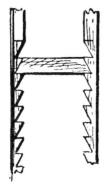

Fig. 216.—Rack for Movable Shelves.

As long bookcase shelves are seldom advisable on account of the weight they have to bear, and consequent tendency to bend, it is usual to make the case when over 4 ft. or 4 ft. 6 ins. long with upright divisions, and often with a break or projecting front in the centre, as in Fig. 217. In this there are three independent sets of shelves. It may be suggested that when a top is of this formation, as it often is in other things besides book-

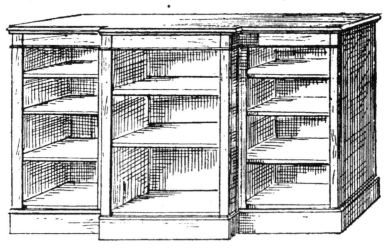

Fig. 217.—Dwarf Bookcase with Break Front.

cases, the best way is to stick the projecting piece on where it is wanted. This is a better way than making the whole of the top to the full width, and then cutting away the spaces at the ends.

Fig. 218.—Secretary Bookcase.

Bookcases and cupboards generally are often made with sliding instead of hinged doors, one sliding behind the other. This arrangement is convenient sometimes, but with glass doors, which are usually found in bookcases, it must not be forgotten that it is not so easy to clean the insides as when they open outwards.

Taller bookcases are generally made in two parts, the lower somewhat deeper from back to front than the upper, which alone has glass doors. Those in the lower or cupboard portion are seldom of anything but wood. Drawers are often added in the lower portion above the doors. As with dwarf bookcases, these higher ones should seldom be made more than 4 ft. 6 ins. wide without a division, or, if very large, more than one.

Instead of having plain drawers, or none at all, the lower part of a bookcase is often fitted for writing purposes, either with a cylinder-fall or with a special arrangement as shown in Fig. 218. In this the front of the writing part folds against inner ends, which can be pulled out and pushed back as occasion requires. The flap or lid is kept up by a spring catch on each side.

Nests of pigeon-holes for containing letters and documents are among the things which the cabinet-maker is sometimes called on for. Their construction calls for no special remarks beyond saying that they are often made with folding shutter fronts constructed like the flexible falls already mentioned for writing-tables.

CHAPTER XX.

SIDEBOARDS AND CABINETS.

Ordinary Arrangement of Sideboards—Fixing of Back—Cabinets—Music Cabinets.

PROBABLY in no articles of furniture is there more variety in design and consequent modifications of construction than in sideboards and cabinets. As some may wonder at these two being classed together, it may be well to explain that so far as the work of the cabinet-maker is concerned they are practically the same in construction, the principal difference otherwise between them is that the sideboard is more massive than the cabinet and less ornate. Of the actual construction it seems almost unnecessary to say anything, as there is little that has not been mentioned elsewhere, and it may reasonably be supposed that those for whom this book is primarily intended will seldom or never be called on to make anything elaborate till they have acquired the requisite amount of skill otherwise than by its aid. A few remarks, however, may serve as a guide and prevent any one being utterly at a loss with regard to special fittings.

The usual sideboard has pedestals at each end, the centre being either open or enclosed wholly or in part, according to the requirements. In one pedestal or cupboard, generally that on the right, a cellarette for holding wine decanters, &c., is usually fitted. It may either take the form of a deep or box drawer, or of a shallow sliding tray, somewhat like those for wardrobes. In either case it is partitioned and lined usually with zinc, this being done by a zinc-worker and not by the cabinet-maker. The top is usually moulded with a heavy lining, so as to give it a massive appearance, and as sideboard tops are among the widest pieces of wood the cabinet-maker requires to use, it is necessary to take ample precautions in the manner already indicated for allowing for shrinkage. The fitting of the back may cause some trouble. When it is composed partly of cupboards or shelves, with supports in front coming down to the top, it may be simply placed on this and secured by screws driven into the frame from underneath. The top, it will be understood, generally projects a little behind to allow of the skirting-board of the room. If it did not do so the back could not rest against the wall as it generally should, unless it were screwed on behind the back edge of the top, which in most instances would be an unworkmanlike proceeding.

An additional support, and a very necessary one when the back is a plain frame without cupboards or shelves, is to make the ends of the framing sufficiently long to extend part way down behind the sideboard, to the ends of which they are screwed. To allow of these pieces being fitted in a sightly manner, part of the back of edge of the top is usually cut away, for it is rarely that a back of either cabinet or sideboard comes to the extreme end of the top. In these, however, and many other details the exact arrangement must depend almost entirely on the design, for a method that would be suitable for one would be quite inapplicable to another. The great thing to be aimed at is stability and, perhaps equally important, neat workmanship.

Ordinary cabinets, as has been said, are very like light sideboards, and beyond saying that the prevailing fashion is to have them inlaid with marquetry, and consequently veneered, no special remarks about them can be necessary.

The music-cabinet is generally only a small carcase fitted with shelves, drawers, and upright partitions sepa-

Fig. 219.—Music-cabinet.

rately or in combination. Fig. 219 represents one of rather more imposing appearance than usual, and though it is mentioned last it may very appropriately, as an exercise of carcase work, be the first which the young cabinet-maker undertakes.

In conclusion, it merely remains to be said that many articles of furniture have necessarily been left unmentioned, and that the present volume has been mainly occupied in describing the groundwork or elements of cabinet-making. Those who understand these and the principles which have been explained will, provided they possess the necessary amount of manual skill, have little difficulty in devising or making soundly constructed workmanlike furniture.

It may be added that it is in contemplation to supplement this volume by others in which articles of furniture of good useful design will be illustrated and their construction minutely explained. Till they appear, the novice will have enough to study in the present one ; and in the meantime I must take leave of my readers.

INDEX.

Adam Bros., 27
Advance in modern construction, 37
Alterations in decorative details, 32
Amateur and professional work, 9
Antique furniture, 15
Architects' designs, 29
Arris, 172
Art furniture, 28, 30; principles, 30
Ash, 44; Hungarian, 44

Ball catch, Bale's, 243
Baywood, 41
Beaded edges, 174
Beading, inlaid, 176
Beads, stopped, 175
Bearers, top, 158; drawer, 158
Bedroom furniture, 279
Beech, 45
Benches, 120; Britannia Co.'s, 122; German, 122
Bench hook, 111
Bevelled panels, 170
Birch, 45
Bits, 93, 139; boring with, 139
Bit guage, 96, 151
Black and gold decoration, 32
Bookcase, dwarf, 309; secretary, 311; shelving, 310
Bow saw, 78
Brace, 93, 139
Brackets, 263
Bradawl, 92
Brads, 66
Brass handles, 249; work, 238
Bureau, 308

Cabine-makers' work, 24
Cabinets, 313; music, 315
Cabinet shops in 1669, 24

Carving, 20; tools, 145
Castors, 244; dining-table, 245; direct bearing, 245; iron plate, 245; paw, 246; rims, 244; screw, 244; socket, 244
Causes of change in style, 32; of bad furniture, 36
Cedar, 42; pencil, 42
Centre bit, 95
Chair work, 13
Chest for tools, 124; and bench combined, 125; of drawers, 294; dressing, 295
Chippendale, 17; style, 25
Chisels, 89
Circular saw, 142
Clamping ends, 153
Clocks, old, 15
Common furniture, 6
Compasses, 100
Construction, mistakes in, 35
Contraction of wood, 57
Convex mirrors, 203
Cork rubbers, 68, 103
Cornice mouldings, 180
Cornices, 192; fastening, 194
Cramps, 104, 107; temporary, 110.
Cutting gauges, 96
Cutting up boards, 147

Deal, 46
Design, improvements in, 28; invention in, 34; principles of, 212
Designers, furniture, 29
Designing, hints on, 34, 209
Desk slopes, 302
Dimensions of furniture, 210
Doors, 190; panels for, 190; sliding, 312; stiles for, 190

Dovetail guide, 143; saw, 77
Dovetailed bearers, 157
Dovetailing, 154
Dowel plate, 67
Dowelled joint, 162
Dowels, 66
Drawer backs, 184; bearers, 187; bottom slips, 184; bottoms, grooves for, 186; fronts, 183, 189; guides, 188; runners, 187; sides, 184
Drawers, 183
Drawing, 213
Drawings, sectional, 216; working, 214
Dry wood, 53
Drying boards, 55
Dust-boards, 188

'Early English' style, 31
'Eastlake' style, 31
Eastlake's *Hints on Household Taste*, 31
Edge joints, 148
Escutcheon plates, 249
Escutcheons, thread, 249

Facing, 157, 182
Figure in wood, 53
Files, 105
Fillister, 3
Finger joint, 265
Firmer chisels, 89
Flattening boards, 56
Flush bolts, 243
Flutes, 176
French nails, 66
Fret machine, 144
Fretwork, 32
Furniture, adaptation of, 18; dealers in antique, 22; early nineteenth century, 27; Elizabethan, 19; old English, 18; sham antique, 15

Gauges, 96
Gimlets, 92

Glass, bevelling of, 201; buying, 198, 201; fitting silvered, 203; fitting transparent, 207; flaws in, 199; measurement of, 197, 201; 203; plate, 196; sheet, 196; shelves, 207; silvering, 202; use of looking, 197; varieties, 198
Glass-paper, 68
Glue, British, 60; colour of, 60; deterioration of, 61; Foreign, 60; Lepage's fish, 63; necessity for, 59; preparation, 60; preservation of, 62; quality of, 59; using, 62
Glue brush, 63
Glue pot, 60; substitutes for, 63
Glued edged joint, 148
Gouges, 90
Grinding plane irons, 85
Grindstone, 103
Grooving, cross grain, 170; for dovetailed bearers, 171; V-shaped, 173

Hall stands, 211
Halved joint, 163; mitred, 164
Hammer, 99
Handsaw, 75; screws, 103
Heppelwhite, 17, 26
Hinges, backflap, 248; butt, 246; card-table, 248; centre, 249; desk, 249; fitting, 247; piano, 249; plates, 249; screen, 249
Holdfast, 144
Hollow planes, 89

Jack plane, 83; using, 136
Joiner's furniture, 2
Jointer plane, 138
Joints, 147; dowelled, 151; plain glued, 148

Knuckle joint, 263

Lap dovetail, 156
Lathe, Britannia Co.'s, 143; cheap, 144
Lining up ends, 168; tops, 165

INDEX.

Locks, 240; box, 241; cut cupboard, 241; desk, 242; piano, 243; straight cupboard, 242; till, 240; wardrobe, 242

Mahogany, 40; decline of, 32; early, furniture, 25; introduction of, 24; Spanish, 41
Mallet, 100
Manwaring, 26
Marking awl, 101; gauge, 96
Marquetry, 33
Mitre block, 113; box, 114; dovetail, 156; shoot, 112; square, 98
Mitred corner keys, 159
Moisture in wood,
Mortise, 160; chisels, 90; gauge, 96; and tenon joint, 159
Mouldings, 177; for rabbets, 191
Munting, 186

Nails, screw, 64
Needlepoints, 66
New Zealand woods, 48

Oak, 43; American, 43; colour of old, 21; English, 43; for fumigation, 43; pollard, 43
Ogee moulding, 178
Oilstone, 101; 229
Old woman's tooth, 89
Ovolo moulding, 178

Panels, 181
Paring chisels, 89
Pedestals, bedroom, 300
Pepys, quotation from, 23
Pigeon holes, nests of, 312
Pincers, 98
Pine, 45; Californian red, 47; Kauri, 48; pitch, 46
Plain dovetail, 154
Planes, 79
Plane irons, 82, 84, 136
Planing, 135
Pliers, 98
Plinths, 192
Plough plane, 88

Practice necessary, 131
Punch, 100

'Queen Anne' furniture, 24

Rabbets, 169; plane, 3, 87
Rasp, 105
Rosewood, 33, 44
Round planes, 89
Router, 116; irons, 119; use of, 116, 118
Rule, 100
Rule joint, 269

Satinwood, 45
Saws, 74
Saw teeth, 76; setting, 77; sharpening, 77
Sawing, 132; cross cut, 134; rip, 134; trestles, 132, 135
Scraper, 101; sharpener, 101; sharpening, 130; using, 139
Scratch, 116; use of, 118; irons, 119
Screws, brass, 65; for wood, 64
Screw box and tap, 108
Screwdrivers, 96
Screws, dowel, 64; sizes of, 65
Seasoned wood, 53
Sequoia, 47
Sham solidity, 31
Shell gimlet, 93
Sheraton, 27
Shrinkage of wood, 53
Shooting board, 111, 137
Sideboards, 313; fittings of, 313; Tudor, 16
Skill requires practice, 11
Sliding bevel, 99
Smoothing plane, 86; using, 137
Spindle rails, 257
Spokeshave, 92
Spring catches, 243
Square, wooden, 115; legs, 258
Squaring up boards, 146
Squares, 98
Stopping, 68
Stop chamfering, 172

INDEX.

Straight edges, 115, 137
Strength, 38
Strengthening plain joint, 150
Stringing, inlaid, 176
Subdivision of work, 5
Substance of parts, 212
Superficial measurement, 51
Superiority of modern work, 6
Swinging legs,

Tables, 253; card, 272; cylinder writing, 305; dining, 274; double Sutherland, 267; flap, 262; folding, 272; half-register writing, 303; kitchen, 259; lining tops of writing, 261; pedestal toilet, 299; pedestal writing, 301; register writing, 303; Sutherland, 266; small round, 259; slides of dining, 275; leaves of dining, 277; toilet, 290
Table tops, fastening, 256; lining, 262; shrinkage, 256
Tacks, 66
Talbot, Bruce J., 29
Tenon, 160; double, 161; foxed, 162; stub or stump, 161; with haunch, 162
Tenon saw, 77
Test for sharpness, 128
Thought necessary, 12
Thumb moulding, 178
Timber, buying, 48; thickness, 52; measurement, 51; not detailed, 48
Toilet glasses, 294
Tongued joint, 152
Tool grinding, 128
Tools, care of, 72; condition of, 71; list of, 73; new forms of, 126;

purchasing, 72; selection of, 71; sharpening, 127
Toothed plane, 88
Towel rails, 300
Trencher for grooving, 171
Try planes, 85, 138
Try square, 98
Twist gimlet, 93

Variety of furniture, 39
Veneer hammer, 233
Veneers, laying, 229; laying light, 230; utility of, 222
Veneered work, 31, 35, 221; blisters on, 231, 235; cleaning, 231
Veneering cauls, 224; flexible, 236
Veneering, cost of, 221; on end grain, 235; with hammer, 232
Veneers, burr, 223, 236; ground for, 227; inlaid, 236; preparation of, 226; sorts of, 224

Walnut, 43; American, 43; burr, 44; Italian, 44; satin, 44
Wardrobes, 279; Beaconsfield, 284; fittings, 288; hanging, 280; hanging and drawer, 282; with straight ends, 283; three door, 286; trays, 287
Washstands, 290
Washstand, marble, 295; pedestal, 298; with pedestal, 297; tile backs, 296
Waste in timber, 58
Whitewood, American, 47
Winding sticks, 116; using, 141
Wire nails, 66
Wooden cramp, 107

Xylonite, 177

LONDON:
Printed by SRANGEWAYS & SONS, Tower Street, Cambridge Circus, W.C

MOSELEY & SON,
323 HIGH HOLBORN, W.C.
Near CHANCERY LANE (formerly of Covent Garden),

Makers of Saws, Planes, Lathes, Carving Tools, Fret-Machines, and Fret-Workers' Tools,
AND
TOOLS OF EVERY DESCRIPTION FOR MECHANICS AND AMATEURS.
Contractors to London School Board, Technical Schools, &c., &c., &c.
WE STOCK THE NOTED 'SHAMROCK' BRAND OF TOOLS.

OUR SMOOTHING

CELEBRATED PLANES.

See Preface.

	EACH		EACH
2 in. C. S. Double Iron Plane, warranted,	3/4	2½ in. C. S. Double Iron Plane, warranted,	5/0
2¼ in. ditto ditto	3/6	2¼ in. ditto, with moving iron fronts ...	8/6
2½ in. ditto ditto	4/0	2½ in. Smoothing Planes, iron sole ...	10/0
2¾ in. ditto ditto	4/6	Toothing Planes	3/0

JACK PLANE.

Jack Planes, 17 in. long, 2¼ iron	5/- each.
Trying do. 22 in. do. 2¼ do.	6/6 ,,
Badger Planes	8/6 ,,
Panel Planes ... ... 6/-, slip	7/- ,,
Jointers' do. 24 in., 7/-; 26 in., 7/6; 28 in., 8/-; 30 in.	8/6 ,,

ORDERS OVER 10/- CARRIAGE PAID.

NOTE.—Our New Illustrated 200-page Catalogue is now ready, containing 700 Illustrations of all the latest Improved Tools for Carpenters, Joiners, Engineers, and all Metal Workers, Carvers, Fret-workers, &c. By Post, 6d.

Y

TO WOODWORKERS.

BEFORE DECIDING WHERE TO BUY

GAS, STEAM, OR PETROLEUM ENGINES,

Send for our Catalogue, Six Stamps, or Twopence for our Monthly Register of

Second-hand Builders' and Contractors' Tools, Plant, Petroleum Engines, &c.

THREE THOUSAND LOTS.

BRITANNIA COMPANY, 100 HOUNDSDITCH, LONDON.

All Letters Britannia Tool Works, Colchester.

THE SECOND EDITION REVISED IS NOW READY.

WOOD-CARVING, By CHARLES G. LELAND,
F.R.L.S., M.A. Fcap. 4to. With numerous Illustrations. 5s.

'A very useful book.'—Mr. W. H. HOWARD, *Secretary to the Institute of British Wood Carvers, and Instructor at King's College, London.*

'A splendid help for Amateurs and those beginning the trade. Without exception, it is the best book I have read at present.'—Mr. T. J. PERRIN, *Society of Arts Medallist, Instructor in Wood Carving at the People's Palace.*

'I consider it the best manual I have seen.'—Miss HODGSON, *Instructor in Wood Carving at Manchester Technical School.*

'Such patient, explicit, step-by-step teaching as Mr. Leland's is indeed the only road to excellence.'—*Saturday Review.*

'An excellent manual.'—*Morning Post.*

'An admirable little book.'—*Builder.*

'Far the most thorough work on the subject that has appeared.'—*St. James's Gazette*

'A thoroughly practical manual.'—*Speaker.*

'It treats of wood-carving very clearly and practically.'—*Spectator.*

'It is ingeniously progressive, and is written with admirable clearness.'—*National Observer.*

'A clearly written, beautifully and effectively illustrated, and well-printed guide.'—*Work.*

LONDON: WHITTAKER & Co., PATERNOSTER SQ., E.C.

WHITTAKER'S
PRACTICAL HANDBOOKS.

FULL LISTS FREE UPON APPLICATION.

The Practical Telephone Handbook and Guide to Telephonic Exchange. By J. POOLE (Wh. Sc. 1875), late Chief Electrician to the Lancashire and Cheshire Telephonic Exchange Company, Manchester. 300 pages. 227 Illustrations. 3s. 6d.

The First Book of Electricity and Magnetism By W. PERREN MAYCOCK, M.Inst.E.E. With 85 Illustrations. Cloth, 2s. 6d.

The Optics of Photography and Photo-graphic Lenses. By J. TRAILL TAYLOR, Editor of the *British Journal of Photography*. 244 pages. 68 Illustrations.

The Art and Craft of Cabinet-Making. By D. DENNING. 320 pages. 219 Illustrations.

The Electro-platers' Handbook. By G. E. BONNEY. 208 pages. 62 Illustrations, and full Index. 3s.

Metal Turning. By a FOREMAN PATTERN MAKER. With 81 Illustrations, and Index. 4s.

Practical Ironfounding. By the Author of *Pattern Making, Lockwood's Dictionary of Mechanical Engineering Terms*, etc. 212 pages. 109 Illustrations, and Index. 4s.

Electro-Motors: How Made and How Used. By S. R. BOTTONE. 166 pages. 64 Illustrations, and Index. Second Edition, Revised. 3s.

Electrical Instrument Making. By S. R. BOTTONE. Fourth Edition, Revised. 202 pages. 65 Illustrations, and Index. 3s.

Electric Bells. By S. R. BOTTONE. 204 pages. 99 Illustrations. Third Edition, Revised. 3s.

Electric Light Installations and the Man-agement of Accumulators. By SIR DAVID SALOMONS, Bart. With 438 pages and 106 Illustrations. Sixth Edition.

'Contains a vast amount of really useful information.'—*Electrical Review.*
'From a practical point of view the work is an excellent book of reference.'
—*Electrician.*

Electric Influence Machines. By J. GRAY, B.Sc. 252 pages and 89 Illustrations. 4s. 6d.

Electricity in our Homes and Workshops. By SYDNEY F. WALKER, M.I.E.E., M.I.M.E., Assoc. M. Inst. C.E. Second Edition. With 320 pages and 127 Illustrations. 5s.

Foden's Mechanical Tables. Fifth Edition. Cloth, 1s. 6d.

Wood Carving. By C. G. LELAND. With 86 Illustrations. Many of them full page. Foolscap 4to. 170 pages. 5s.

'A very useful book.'—Mr. W. H. HOWARD, *Secretary to the Institute of British Wood Carvers, and Instructor at King's College, London.*

'A splendid help for Amateurs and those beginning the trade. Without exception, it is the best book I have read at present.'—Mr. T. J. PERRIN, *Society of Arts Medallist Instructor in Wood Carving at the People's Palace.*

'I consider it the best manual I have seen.'—Miss HODGSON, *Instructor in Wood Carving at Manchester Technical School.*

Drawing and Designing. By C. G. LELAND. 1s. Sewed; 1s. 6d. Cloth.

'Full of valuable practical suggestions for beginners.'—*Scotsman.*
'The book deserves the widest success.'—*Scottish Leader.*

Pictorial Astronomy. By G. F. CHAMBERS, F.R.A.S. Square crown 8vo. 4s.

Light. By SIR H. TRUEMAN WOOD, M.A. Square crown 8vo. 3s.

The Plant World: Its Past, Present, and Future. By G. MASSEE, of Kew Gardens, with many Illustrations. Square crown 8vo. 3s. 6d.

Practical Education. A Work on Preparing the Memory, Developing Quickness of Perception, and Training the Constructive Faculties. By CHARLES G. LELAND.

The Electric Transmission of Energy. By GISBERT KAPP, C.E. 7s. 6d.

Conversion of Heat into Work. By W. ANDERSON, F.R.S., D.C.L. 6s.

Hydraulic Motors. By G. R. BODMIN, Assoc. M. Inst. C.E. 14s.

The Telephone. By W. H. PREECE, F.R.S., and Dr. J. MAICE. 12s. 6d.

Alternating Currents of Electricity. By T. H. BLAKESLEY, M.A., M. Inst. C. E. 5s.

Colour in Woven Design. By Professor ROBERTS BEAUMONT. 21s.

A Treatise on Manures. By Dr. A. B. GRIFFITHS. 7s. 6d.

The Alkali Maker's Handbook. By Profs. Dr. LUNGE and Dr. HURTER. 10s. 6d.

WHITTAKER & Co., Paternoster Square   LONDON, E.C.

Hertz' Work, Lodge, 2s. 6d. net.
Hewitt and Pope's Elem. Chemistry, 9d. net.
Highways Management, Hooley, 1s.
—— Bridges, Silcock.
Hobbs' Electrical Arithmetic, 1s.
Hoblyn's Medical Dictionary, 10s. 6d.
Holtzapffel's Turning, 5 vols., 5l. 9s.
Hooley's Highways 1s.
Hooper's Physician's Vade Mecum, 12s. 6d.
Hopkinson's Dynamo Machinery, 5s.
Horner's Mechanical Works.
Hospitalier's Polyphased Alternating Currents, 3s. 6d.
Houston's Electrical Terms, 21s.
—— Electricity Primers, 3 vols.
Hurter's Alkali Makers' Handbook, 10s. 6d.
Hutton's Mathematical Tables, 12s.
Hydraulic Motors, Bodmer, 14s.

Imray and Biggs' Mechanical Engineering, 3s. 6d.
Incandescent Lamp, Ram, 7s. 6d.
Induction Coils, Bonney, 3s.
Industrial Instruction, Seidel, 2s. 6d.
Inventions, How to Patent, 2s. 6d. net.
Iron Analysis, Arnold, 10s. 6d.
—— Analysis, Blair, 18s.
—— and Steel, Skelton, 5s.
Ironfounding, 4s.

Jack's Cooking, 2s.
—— Laundry Work, 2s.
Jacobi's Printer's Handbook, 5s.
Jones' Refuse Destructors, 5s.
Jukes-Browne's Geology, 2s. 6d.

Kapp's Alternating Currents, 4s. 6d.
—— Dynamos, &c., 10s. 6d.
—— Electric Transmission of Energy, 10s. 6d.
—— Transformers, 6s.
Kennedy's Electric Lamps, 2s. 6d.
Kennelly's Electrical Notes, 6s. 6d.
Kilgour's Electrical Formulæ, 7s. 6d.
—— Electrical Distribution, 10s. 6d.
Kingdon's Applied Magnetism, 7s. 6d.
Klindworth's Stuttering, 12s. 6d.

Laundry Work, Jack, 2s.

Leland's Wood-carving, 5s. Metal Work, 5s. Leather Work, 5s. Drawing and Designing, 1s. and 1s. 6d. Practical Education, 6s.
Lens Work for Amateurs, Orford, 3s.
Lenses, Photographic, Traill Taylor, 3s. 6d.
Library of Arts, Sciences, &c.
—— of Great Industries.
—— of Popular Science, 2s. 6d. per vol.
Light, Sir H. T. Wood, 2s. 6d.
Lightning Conductors, Lodge, 15s.
Lockwood's Telephonists, 4s. 6d.
Lloyd's Mine Manager, 1s. 6d.
Locomotives, Cooke, 7s. 6d.
—— Reynolds, 2s. 6d.
Lodge's Lightning Conductors, 15s.
—— Hertz, 2s. 6d. net.
Lukin's Turning Lathes, 3s.
—— Screws, 3s.
Lunge and Hurter's Alkali Makers' Handbook, 10s. 6d.

Maclean's Physical Units, 2s. 6d.
Maginnis' Atlantic Ferry, 7s. 6d. and 2s. 6d.
Magnetic Induction, Ewing, 10s. 6d.
Magnetism, Kingdon, 7s. 6d.
Manchester Ship Canal, 3s. 6d.
Manual Instruction and Training.
Manures, Griffiths, 7s. 6d.
Marine Engineering, Maw, 3l.
Marshall's Cakes, 1s.
Martin's Structures, 4s.
Mason's Sanitation.
Massee's, The Plant World, 2s. 6d.
Mathematical Tables, 12s.
Maver's Quadruplex, 6s. 6d.
Maw's Marine Engineering, 3l.
May's Ballooning, 2s. 6d.
—— Belting Table, 2s. 6d.
—— Electric Light Plant, 2s. 6d.
Maycock's Electricity and Magnetism, 2s. 6d.
—— Electric Lighting, 6s.
Mechanical Tables, 1s. 6d.
—— Eng., Imray and Biggs, 3s. 6d.
Medical Terms, Hoblyn, 10s. 6d.
Merrill's Electric Lighting Specifications, 6s.
Metal Turning, 4s.

Metallurgy, Gold, 31s. 6d. Silver and Mercury, 31s. 6d. Egleston.
Metric System, Wagstaff, 1s. 6d.
Metric Measures, Born, 3s.
Middleton's Surveying, 4s. 6d.
Mill Work, Sutcliffe, 21s.
Mine Manager, Lloyd, 1s. 6d.
—— Ventilation, Tate, 6d.
—— —— Halbaum, 1s.
Mines (Fiery) Management.
Mining Arithmetic, W. Tate, 6d.
—— Examinations, Davies, 2 parts, 1s. 6d. each. [6d.
—— Student's Examples, W. Tate,
Mining Students' Handbooks, 4 parts, 6d. each.
Mineralogy, Hatch, 2s. 6d.
Minstrelsie, English, 8 vols., 4l. Scots, 6 vols., 2l. 11s. Cambrian, 2l. 11s.
Mitton's Fiery Mines, 1s.
Model Steam Engine Making, 10s. 6d.

Nadiéine, Drainage, 1s.
Naval Tactics, Fitzgerald, 1s.
Niblett's Electricity, 2s. 6d.
—— Secondary Batteries, 5s.
Nicholl's Agricultural Engineering in India, 3s. 6d.
Noll's Wiring, 6s.

Optical Instruments, Oxford, 2s. 6d.
Optics of Photography, 3s. 6d.
Orford's Lens Work, 3s.
—— Modern Optical Instruments,
Ozone, Andreoli, 2s. 6d. [2s. 6d.

Parkhurst's Dynamo Building, 4s. 6d.
Parshall's Armature Winding, 30s.
Patenting Inventions, 2s. 6d. net.
Pattern Making, 3s. 6d.
Petroleum, Boyd, 2s.
Philosophical Mag., *Monthly*, 2s. 6d. net.
Photographic Lenses, Taylor, 3s. 6d.
Physician's V.M., 12s. 6d.
Physical Units, Maclean, 2s. 6d.
Pickworth's Slide Rule, 2s.
Plant World, Massee, 2s. 6d.
Planté's Electric Storage, 12s.
Ponce de Leon's Spanish Technological Dictionary. Vol. I., 36s.

Practical Education, 6s.
Preece's Telephony, 15s.
Price's Hoblyn's Dictionary, 10s. 6d.
Primers of Electricity, 3d. each.
Printer's Handbook, Jacobi, 5s.
Printing, Southward, 10s.
Pruning, Des Cars, 2s. 6d.
Public Arms, 3l. 3s.

Quadruplex, Maver, 6s. 6d.
Questions in Typography, 6d.

Railway Management, Findlay, 7s. 6d.
—— Material Inspection, Bodmer.
—— Pennsylvania, 52s. 6d.
Ram's Incandescent Lamp, 7s. 6d.
Reckenzaun's Electric Traction, 10s. 6d.
Refuse Destructors, Jones, 5s.
Repoussé, Leland, 5s.
Reynold's Locomotive, 2s. 6d.
Röntgen's X Rays, 9d. net.
Russell's Electric Cables, 7s. 6d.
—— T. C.'s Handbook to Electric Lighting, 1s.

Salomons' Electric Light Installations, 6s.
—— Electric Light. Vol. I., Accumulators, 5s. Vol. II., Apparatus, 7s. 6d. Vol. III., Applications, 5s.
Sanitary Drainage, 1s.
Sanitation, Mason.
Savage's Sewage Disposal, 5s.
Scholl's Phraseological Dictionary, Eng., Ger., Fr., and Span., 21s.
Scots Minstrelsie, 2l. 11s.
Screws and Screw Making, 3s.
Screw Propulsion, 15s.
Segundo's Dom. Elec. Lighting, 1s.
Seidel's Industrial Instruction, 2s. 6d.
Sewage Disposal, Savage, 5s.
—— Treatment, 6s.
Shafting and Gearing, Bale, 2s. 6d.
Ships' Resistance, 15s.
Silcock's Highway Bridges.
Skelton's Iron and Steel, 5s.
Slater's Sewage Treatment, 6s.
Slide Rule, Pickworth, 2s.
Sloyd, English, 7s. 6d.
Smith's Cable Traction, 5s.

Printed in Great Britain
by Amazon